ROUTLEDGE LIBRARY EDTIONS:
GLOBAL TRANSPORT PLANNING

I0127576

Volume 17

CITY CENTRE PLANNING AND PUBLIC TRANSPORT

CITY CENTRE PLANNING AND PUBLIC TRANSPORT

Case Studies from Britain, West Germany and France

BARRY J. SIMPSON

Routledge
Taylor & Francis Group

LONDON AND NEW YORK

First published in 1988 by Van Nostrand Reinhold (UK) Co Ltd

This edition first published in 2021
by Routledge
2 Park Square, Milton Park, Abingdon, Oxon OX14 4RN

and by Routledge
52 Vanderbilt Avenue, New York, NY 10017

Routledge is an imprint of the Taylor & Francis Group, an informa business

British Library Cataloguing in Publication Data
A catalogue record for this book is available from the British Library

ISBN: 978-0-367-69870-6 (Set)
ISBN: 978-1-00-316032-8 (Set) (ebk)
ISBN: 978-0-367-74232-4 (Volume 17) (hbk)
ISBN: 978-0-367-74252-2 (Volume 17) (pbk)
ISBN: 978-1-00-315677-2 (Volume 17) (ebk)

Publisher's Note
The publisher has gone to great lengths to ensure the quality of this reprint but points out that some imperfections in the original copies may be apparent.

Disclaimer
The publisher has made every effort to trace copyright holders and would welcome correspondence from those they have been unable to trace.

City Centre Planning and Public Transport

Case studies from Britain, West Germany and France

Barry J. Simpson

VNR
UK

Van Nostrand Reinhold (UK) Co. Ltd

First published in 1988 by
Van Nostrand Reinhold (UK) Co. Ltd.
Molly Millars Lane, Wokingham, Berkshire, England

Typeset in Century Schoolbook 10/12pt by
Columns of Reading
Printed and bound in Great Britain by
T.J. Press (Padstow) Ltd., Padstow, Cornwall

British Library Cataloguing in Publication Data

Simpson, Barry J.
    City centre planning and public transport:
    case studies from Britain, West Germany
    and France.
    1. Urban transportation – Europe –
    Planning – Case studies
    I. Title
    711'.7'094   HE4704

ISBN 0–7476–0006–6

Contents

Preface

Loss of 'bright lights' speciality shopping, offices moving to the suburbs or further afield, closing of small scale industries, loss of housing, street violence – these are amongst the issues facing many West European cities. For a long time, city centres were taken for granted. Some municipalities assumed they would always be there, vibrant and prosperous as ever, to house all the leisure, cultural, employment and other wants and to bring in revenue to the municipal coffers. Some city authorities even planned decline.

The effects of the rise in private car use from the 1950s onwards have taken a long time to become evident. The car favours dispersal of shopping, leisure and commerce to the suburban centres, out-of-town sites and cheap sites in the inner city not far from the centre. These have been rendered accessible to large numbers of people with cars. These sites can also offer better parking and access for the car than many city centres. The car has also taken away many public transport customers and contributed to the weakening of city centres which have long relied on it.

Can light rapid transit save the city centre? How far is the private car to be accommodated in the city centre? Most large German cities have had *U-Bahn* or light rapid transit since the late 1960s or early 1970s, *métros* have been opened in Marseille, Lyon, Tyne & Wear, Lille and Nantes since 1977, and several other large British cities such as Manchester, Sheffield and Birmingham are thinking about it. What does the experience in West Germany and France have to offer? Should car parking be further restricted so as to divert travellers to public transport, or is the risk of diverting them out of the city centre too great? Instead, should they be persuaded to use public transport by making it more attractive with large investments in light rapid transit? Will the investment in light rapid transit (and roads and parking) in this high spending option be justified by the returns, economic and otherwise, from city centre recovery? Can pedestrianization and other environmental improvements be just as effective? These are some of the most fundamental questions facing city-centre planners and public transport planners today. They can be approached only gradually and even then only with the big assumption that what may have been a justified investment in the past would still be justified if made at present.

During the writing of this book I have received help of many kinds. Above all, my thanks go to the Nuffield Foundation for generously funding research projects at Aston University and to their Deputy Director, Miss Patricia Thomas, for her efficient administration of these. Since 1978 they have supported a series of research projects I have carried out at Aston University and this book would not have been written without their support.

Many officials in this country, West Germany and France have generously given up their time for discussion and have supplied me with information. The following deserve special mention:

M. Jean–Pierre Aldeguer, Université de Lyon.

Frau Rosemary Holley, Städtepartnerschaften, Magistrat der Stadt Frankfurt am Main and the City of Frankfurt for their generous hospitality during my visit there in April 1985.

M. Herbulot, Directeur, and MM. Tulasne, Joannon and Romano of the Agence d'Urbanisme de l'Agglomération Marseillaise and the City of Marseille for their generous hospitality during my visit there in March 1986.

M. Courcoul of the Régie des Transports de Marseille.

M Clavier, Directeur du Société du Métro de Marseille.

M. Coste, Secretaire Général Adjoint de l'Office de Coordination des Transports, de la Circulation et du Stationnement de Marseille.

Mr F. G. Simons, Mr E. Godward, Mr P. Evans and Mr Martin Watkins of the West Midlands Passenger Transport Executive.

Professor Dr-Ing P. A. Mäcke, Direktor und Ordinarius, Institut für Stadtbauwesen, Rheinische Westfälische Technische Hochschule, Aachen.

Mlle Martine Rivet, Mme Pascale Pernet and Mlle Christiane Dalmais van Straaten, Agence d'Urbanisme de la COURLY, Lyon.

M. Gilbert Lazar, Ecole d'Architecture de Lyon.

Mlle Pierette Dufour, Compagnie du Métro de Lille.

Herr Dr Brandt, Herr Spuck, Herr Baerwolf, Herr Kohler and Herr Erker, Frankfurter Verkehrs- und Tarifverbund.

Herr Kohlberger, Herr Dr Janz, Herr Hausmann, Herr Dr Gehrke, Herr Dr Vaubel and Herr Schmdt of the Ämt für kommunale Gesamtentwicklung und Stadtplanung, Frankfurt am Main.

M. Jean–Pierre Steffen, Chef du Service Marketing, Société Lyonnaise de Transports en Commun.

M. Michel Watel, Service Urbanisme Commercial et Etudes, Chambre de Commerce et d'Industrie de Lyon.

Professor Denys Hinton, Chairman of Redditch Development Corporation and Mr Stuart MacNidder of the Town Planning Department.

Herr Schünemann, Tiefbauamt, Hansestadt Lübeck.

Herr Flathmann and Herr Wenke, Amt für Strassen und Brückenbau, Freie Hansestadt Bremen.

M. P. Dassonville and Melle Ganter, Service des Transports, Communauté Urbaine de Lille.

Ms Sandra Hanafin, GLC Research Library.

Frl Susanne Rohrberg, Studiengesellschaft Nahverkehr mbH Hamburg

M. Pierre Barbaudy, Régie Autonome des Transports Parisiens, Service du Developpement Commercial.

Herr Axel Holst, Bundesministerium für Verkehr, Bonn.

Herr Dipl–Ing Ralf Stock, Allgemeiner Deutscher Automobil–Club eV, Munich.

Herr B. von Sturm, Hamburger Verkehrsverbund.

Mme J. Boller, Société d'Economie Mixte du Métropolitain de l'Agglomération Lyonnaise.

Finally, my thanks are due to Mr Lionel Browne and the staff at Van Nostrand Reinhold for their helpful and efficient administration, and to the staff of the Department of Civil Engineering at Aston University for providing surroundings in which the writing was possible and the recognition which made it worthwhile, and in particular Mrs Joy Atkins and Miss Sally Fulford for their help with the preparation.

B. J. Simpson MSc PhD,
Department of Civil Engineering,
Aston University, Birmingham B4 7ET.

1 Issues

As long as there have been cities, they have been the locations of services for their surrounding catchment areas – serving as markets and places of worship, entertainment and political power, for example. Concentrating these facilities together reduces the distance to be travelled by those who visit all or some of them. Very simply, this is why cities have always had central areas with uses distinguishing them from the rest of the city: a visit to one location gives access to a whole range of trading, entertainment and cultural facilities.

Cities are not only convenient for those travelling to them. They are efficient for those operating in them too. Traders, employers and others benefit from being in the focus of visitors. Because of their large catchment area they are able to operate more efficiently than if they were dispersed, and are able to compete successfully with those offering similar services elsewhere.

City centres therefore offer minimization of travel costs for visitors and maximization of catchment area and efficiency of operation for those in business there or otherwise relying on visitors. These have always been overwhelmingly good reasons for their existence. However, their success has brought with it the seeds of failure. Being attractive locations for traders, businesses and others depending on a large clientèle, taxes and land values in city centres have risen. Other bodies, including local and central government, have muscled in to cream off some of the profits. The businesses themselves have bid against each other to give some of their profits to land-owners and developers. Those to whom a central location was less valuable found it advantageous to move to a lesser centre, especially if they were particularly sensitive to some of the other consequences of high land values (high density and congestion, for example) and not dependent on contact with other city centre users.

Planned new towns, new suburban centres and out-of-town

shopping centres ('centre' being a significant expression) therefore defy the forces which have resulted in city-centre development. They have largely avoided the detrimental side-effects of urban concentration (high land prices, lack of space, congestion) whilst forgoing the benefits of a ready-made clientèle. They have also been an attempt to start centralization afresh with resulting benefits and without the accrued disbenefits of older city centres.

Much of the dispersal of centres took place before the increase of the motor car which has occurred since the early 1950s. The Garden City Movement from the end of the nineteenth century and the suburbanization of many British cities in the early part of this century, aided by the development of tramway networks, are examples. The dispersal induced by the motor car has, in the UK, been restricted, owing to the advent of planning legislation to control all development in 1948, just at the beginning of the period when the car was to increase so dramatically in numbers. The effects of the car in permitting dispersal are much clearer and more widespread in North American cities. The private car has made distance less significant. It has weakened the significance of established city centres as a means of minimizing travel distance and has concentrated users' minds on their disadvantages.

Weakening the city centre by use of the private car has been resisted on social grounds. Half the population of Britain do not have access to a car. A few out-of-town shopping centres have provided special buses free of charge, but a car remains close to being essential to use them. Local authorities have also restricted out-of-town shopping on economic grounds. Understandably they do not wish to see a decline of existing centres and, consequently, loss of rates revenue. However, this seems to be no more than an administrative reason, resulting from the disposition of municipal boundaries and methods of raising tax. Out-of-town shopping centres seem much more likely to result in a transfer rather than a loss of activity and income.

Issues facing city centres today

Cities in much of Western Europe have a lot in common in the present-day issues that they face. A few grand-scale projects, such as Les Halles in Paris and Part Dieu in Lyon, reminiscent of the 1960s and early 1970s, are still under way, but generally there has been a pronounced swing away from large-scale redevelopment to small-scale enhancement and conservation. City centres are now facing decline rather than growth, and consequently there has been a change of emphasis from the planning of development to the management of the environment. Accompanying this change there has been a growing appreciation of the potential of public transport in social, economic and environmental terms and the need to control the escalating cost of such transport, aggravated by competition from the private car.

It is, however, not entirely clear to what extent city-centre development should be planned for, or even how far decline should be arrested. Developing the city centre is often at the expense of inner city districts and results in environmental costs (Centre d'Etudes des Transports Urbains 1978b).

Decline of shopping, offices and commercial activities: decentralization

Planning powers cannot prevent any activity ceasing, but (legally at least) can prevent them starting up elsewhere. Whether this is practical depends on the resolve of local and central government in applying planning policy, bearing in mind that, as in any field, tough policies can sometimes stifle activities setting up altogether. An intending developer who is excessively restricted might decide he is in the wrong city, the wrong country or even the wrong business.

Conversely, on the carrot principle, local authorities can provide infrastructure which will increase the attractiveness of city-centre development. Public transport and roads are perhaps amongst the most potent attractions that a municipality can provide. Certainly a lot of cities have been willing to back up this view with very heavy investment in public rail transport of various kinds.

Because of the range of goods offered, out-of-town shopping centres have been far more of a threat to suburban shopping than to the city centre. Cheap premises in the inner city or on the fringes of the city centre have been significant in catering for durable cash-and-carry goods of a kind traditionally offered in the city centre. Paradoxically, out-of-town office centres seem to have met with less resistance from planning authorities. La Défense in Paris and Croydon in London are large scale examples. On a provincial scale, Five Ways in Birmingham has drawn employment away from the city centre. Perhaps some decentralization is not so bad, considering effects on city centre traffic and the need to satisfy office users' requirements for car parking space, accessibility and rent or price.

These issues are really just a few of the effects of a whole series of processes affecting the urbanization and spatial development of cities. Hall, Thomas, Gracey and Drewett (1973) give a comprehensive account of metropolitan growth processes in Britain since the Town and Country Planning Act 1947, whereas Reichert and Remond (1981a and b) are useful reviews from the French point of view.

Decline and decentralization are certainly seen as a problem by city authorities and their employees. Transfer of activities has received less public comment from receiving authorities. Decline and decentralization are largely a problem seen from a perspective which sees economic success as the prime aim of planning. Whilst it is true that cultural innovation has often come out of the city as a meeting point for ideas, the connection between rateable value and any welfare other than economic seems tenuous. If one of the purposes of planning is to allow for the free development of the individual within society (cf. the West German Basic Law: Chapter 3 below), much of the investment in city centres seems to be orientated towards a narrowly defined range of economic ideals.

Poor physical and social environment; crime against property and person

To some extent these are the accompaniments of success. An attractive city centre will be one of intense activity, liable to congestion and lack of space. They are also the results of changed requirements. Many cities have street patterns of the pre-motor age and a centre suited to a much smaller city. As a city grows, there may be problems in the centre expanding correspondingly.

The alarming increase in street crime in the last decade or so has left many cities unprepared. Many of the designs with secluded underpasses and separation of pedestrians from traffic are totally unsuited to the surveillance that is necessary in an increasingly lawless society. Perhaps designers misread the society they planned for at the time. Whether they did or not, buildings do not change so readily as do social conditions. Research in Frankfurt has suggested that pedestrianization and other traffic restriction reduces some types of street crime, presumably because many petty criminals are car- or motorcycle-borne (Frankfurt am Main, Dezernat Planung, 1977, pp. 12–13).

Certainly designers are not the accomplices of a large part of urban crime. Marseille is not a city of designed seclusion, and yet the visitor is likely to be warned over and over again by residents of its problems of street violence. A visitor who witnesses a knife fight in a large city centre department store by North African immigrants in mid afternoon might be assured that this is not at all uncommon and therefore be sceptical about the adage that crime needs seclusion. On the contrary, it is possible to walk around Hamburg after dusk and see unaccompanied girls and young women walking or cycling along the most secluded paths segregated from the main circulation network. The urban design of many German cities would result in large no-go areas if occupied by a less law-abiding society.

A great deal has been written and said, if rather less done, about the problems of the British inner city. The Census and other statistical sources could be selected at random for any of the large cities and they would tell the same story: the inner areas are worse than average for practically all the indicators of urban deprivation. However, these problems, so clearly expressed recently by Lord Scarman (1986) are not quite the same as those of the city centre. British inner city districts are areas of declining industry and still substantial, if rapidly declining, housing. Unlike the situation in France and Germany, few British city centres have a substantial population.

The city centre differs from the inner city in that it has the role of supporting facilities, activities and commercial prosperity which are unique and which serve the whole city and usually a substantially wider area too. The city centre is one of the principal generators of economic prosperity and cultural well-being of the whole city. However, the distinction between the city centre and the inner city cannot be drawn very tightly. Improvements in access to the centre will obviously affect the neighbouring inner districts. The fear has been expressed in relation to the new French *métros* in Marseille, Lyon and Lille (Chapter 6) that improvements of access through the inner city districts will cause their faltering shopping and entertainment facilities to decline even faster. On the other hand, presumably those with access to the *métro* will experience some increase in welfare if they choose the city centre in preference to local amenities. In French and German cities the distinction between inner city and city centre is in some respects less sharp than in Britain. The city centre is more heavily populated in France and Germany and inner city functions spill over into the city centre to a greater extent than in British cities.

How much restriction should be placed on the private car?

It is quite clear that city centres which were developed before the motor age do not have the roads or parking spaces to accommodate all those who might wish to travel in by car. If city councils attempted to cater for demand, much of the built environment would be destroyed and the price paid in terms of the environment too high. On the other hand, a very restrictive policy towards the car, in terms of restricting parking, pricing or simply allowing roads to become congested and self-regulating, would result in a loss of income to the city centre from shoppers, loss of jobs perhaps and loss of wealth-creating activities. The dilemma facing city councils is where to strike the balance between damage to the built and physical environment on the one hand, and loss of income on the other.

Some large cities have responded by defining a core where there will be some restriction on road traffic surrounded by a ring road (usually not completed) with access to car parking and served by high capacity roads from the suburbs and further afield. In cities such as Birmingham, these restrictions in the core have not been severe. Apart from shopping centres specifically designed for the pedestrian, there are few pedestrianized streets or roads with even limited access and there are almost 10,000 off-street car parking spaces adjacent to the ring road which forms a complete circle around the 80 hectares of the city centre. Access by car was given high priority and this resulted in extensive transformation of the built environment in the late 1960s and early 1970s. A probable contributory factor in giving high priority to the car is the fact that Birmingham has only a weakly developed suburban railway network (5% of local public transport journeys) and no underground, metro or trams. High priority for roads has allowed good access by bus as well as by car.

Other cities, for example Frankfurt am Main, have extensive pedestrianized streets in the core. The *U-Bahn* and *S-Bahn* serve the pedestrian area from below, whilst trams and buses serve it from the surface. As in Birmingham, parking is most easily accessible from the ring road, in this case following the line of the former fortifications.

In deciding where to strike the balance between public and private transport, the significant trade-off is usually between the environment and restricting access. A policy of strict environmental conservation will usually necessitate high capacity public transport. Although this need not restrict access directly, in practice it may well result in the loss of the custom and benefits of a large number of people who are not willing to leave their cars in favour of public transport. On the other hand, in many New World cities which have excellent access by car, the environment seems to have suffered so much that few want to travel there. City centres can be strangled by too much access as well as by too little.

Public transport under threat

The effects of the increase in the numbers of private cars on the demand for public transport began to bite in the mid-1960s, and in Britain, West Germany and France legislation for subsidy was passed between 1967 and 1970.

In a study of 36 medium sized French towns of between 30,000 and 80,000 population in 1975, Minvielle (1985) found that the average level of subsidy rose from 17 francs per person to 55 francs per person from 1975 to 1982 (at constant 1982 prices). In a later study of 101 provincial public transport networks he concluded that usage of public transport *is* sensitive to fares.

Table 1.1 Passenger travel on public transport in Britain (thousand million passenger kilometres).

	1970	1982
bus	53	40
rail	36	31

City centres have generally been the most attractive areas for providing public transport and have been amongst the areas best able to fend off competition from the private car. However, congestion by private transport has slowed down the buses, made them less attractive and contributed to their decline. Urban motorways have competed with local rail transport. They now face increased competition from buses, since deregulation in October 1986 (Farrington 1986).

Several cities seem to have successfully reversed a decline in public transport usage. In Marseille the level of public transport usage in 1985 had returned to that of thirty years earlier, after being halved in the late 1960s. The *métro*, reorganization of bus services, tariff reform and revised marketing strategy have all contributed to this (Chapter 6). In London, allowing some routes to go out to tender seems to have contributed to the increase in usage of public transport between 1982 and 1986 (Chapter 9).

City centre issues and public transport

All forms of public transport, but especially the various forms of railway, can bring in passengers with less environmental cost than can the private car. They can help to realize the benefits of centralization at much less cost than the private car. They can do this for all residents except the severely disabled and the very old, and are not restricted to those who hold a driving licence and have access to a car.

In fact, the objectives of public transport can be classified into three groups – social, economic and environmental. Wherever there is public transport there are all three kinds of benefit, but the emphasis between the three varies from place to place.

Public transport as a social service

Perhaps the most obvious and immediately appealing benefits of public transport are social ones. The daily or weekly bus to the isolated village or the bus to the peripheral estate in the big city are obviously important to many of the less privileged, and give a compelling argument for public transport to politicians and public alike. Schoolchildren, teenagers, housewives, the elderly, the lower paid and many others would suffer a reduction in the quality of life if public transport was significantly reduced. Although 70% of households in France have at least one car and 20% have two, even for professional households more than half do not have access to a car during the daytime (Durand and Pêcheur 1985).

Expenditure on local travel by bus in nearly all West European cities declines a lot, as a proportion of income, with increasing income, and, in some cases, even as the absolute amount spent. Expenditure on travel by train usually follows the reverse pattern. Subsidies to bus fares are therefore progressive with income, and to trains regressive. Although subsidizing train fares may appear to be subsidizing the well-off, it is not as simple as that. Not all train users are wealthy, even though the average income is higher. Deterring train users by cutting subsidies may induce them to travel by car, causing congestion and other nuisance as they travel through the poorer suburbs into town. In any case, having higher incomes on average, train users will pay more taxes. The general taxation system is far more effective as a way of redistributing income than public transport subsidies.

Métros and undergrounds vary a lot in their effects according to income. Even those which extend a long way into the outer suburbs, as in London, have a clientèle generally less wealthy than the suburban railways, whilst those confined mainly to the inner suburbs, such as in Marseille, Lyon and Lille, still cater for those who have higher incomes on average than bus users.

It is common in several West European cities for some groups to get fare reductions. For those who get free travel it is clearly a matter of social policy. For the others it is not so clear whether it is a matter of social concern to these groups or one of economic policy to increase the number of journeys they make. For example, in the West Midlands the maximum off-peak fare is 35p. Is this to subsidize housewives who are more likely to be able to travel off-peak than those travelling to work? Or is it to increase the number of off-peak journeys they make on existing services and at very little cost to the transport authority? As most journeys are between suburb and city centre, those who live beyond the inner suburbs will get most benefit. As the middle and outer suburban dwellers tend to have higher incomes, it is tempting to think the policy in this case is more influenced by economic than social motives.

Public transport as an economic asset

Public transport allows a greater number of people to gain access to the city centre and gives access to many who would otherwise not have it. Amongst the clearest examples of public transport being seen as an economic asset is in cities which have built a metro. Invariably, bus services could do the job more cheaply and with almost equal social and environmental benefits. This seems to be particularly true of the second-tier and smaller cities, such as Marseille, Lyon, Lille or Newcastle, rather than Paris or London. In very large cities, the capacities of the bus networks may be inadequate, but this certainly does not apply to cities of a million or less. In these cities there have been additional environmental benefits (small-scale pedestrianization for example) and no doubt some users such as the elderly find the metros more accommodating than buses, but the real reason for these metros remains very largely economic. They are thought to be more attractive to passengers than buses and to be able to compete more successfully with the private car. The evidence available shows that there is some justification for these beliefs, although none of the recent metros has so far been shown to be an outstanding economic success. In relation to Britain and West Germany, Hall and Hass-Klau (1985) recognized that forces of city decline can be deeper than problems of accessibility and in such cases rail cannot save the city.

The town of Besançon in France, which has a population of 125,000, demonstrated how accessibility by public transport can be improved without railways. In the early 1970s several major works were carried out to improve accessibility to the town centre: a ring road to reduce the amount of private traffic in the centre, bus priority measures including a main axis where other traffic is restricted, pedestrianization, through ticketing on the buses and improvements in information about the buses for users. Between 1973 and 1983 the proportion of journeys to the town centre by bus rose from 21% to 50%. The price of commercial floorspace rose from 4,000 francs to 6,000 francs before the *plan de circulation* to 7,000-10,000 francs afterwards. However, deterioration in access by private transport was accused of being responsible for the reduction in the number of hotels from 17 to 11 (and beds from 604 to 393) (Breton and Lengacher 1985).

Public transport is compatible with environmental improvements

The quality of the environment has been shown to be very significant in influencing the economic prosperity of the city centre. In the great majority of cases, pedestrianization coincides with an increase in shopping turnover. Even neighbouring streets left open to traffic seem to benefit by a pedestrianized street bringing in more customers to the area (Roberts 1981). In future, the significance of the environment seems likely to increase, in shopping streets and in other pedestrianized areas. City centres are turning away from manufacturing and even routine service jobs to activities such as financial services, tourism and entertainment, some of which are even more sensitive to environmental quality.

Public transport, especially underground railways, always cause less environmental damage per person carried than the private car or motorcycle. Underground railways have also aided pedestrianization of the streets above, in cities such as Lille, Lyon and Frankfurt, by giving access without surface traffic. Buses seem to be perceived as a serious intrusion in pedestrian streets. Studies by Stewart (1976) found complete absence of vehicles to be an important asset.

Whether public transport will always result in a significant reduction in the general level of car usage outside pedestrian areas is by no means clear. Whilst studies following the opening of the new French *métros* in Marseille, Lyon (Ministère de l'Environnement et du Cadre du Vie, Ministère des Transports 1979) and Lille (Lille, Communauté Urbaine de 1984a) have shown that some passengers have given up their cars to travel on the *métro*, there is no evidence that the overall level of traffic in the centres of large cities has decreased. The cars released by new *métro* users will very likely be used by other members of their families, perhaps to travel into town. Others may travel into town by car to take up the roadspace released by those travelling with the *métro*. More cars come in to bring up the level of congestion to the maximum tolerable, at which point others are dissuaded from using their cars. This is the same phenomenon which has prevented cities grinding to the halt that was predicted in the 1950s. In smaller cities there is evidence that increased investment in public transport does reduce traffic in the central area. This appears to have been the case in Besançon (125,000 population) for example (le Guénédal 1981).

It is tempting but in many cases wrong to say that increased use of public transport improves the environment. What it does is to bring more people in before the environment sinks to the lowest level tolerable to large numbers of city centre users. It allows some environmental improvement such as pedestrianization without having to face the economic disbenefits of restricted access.

Britain, West Germany and France

Table 1.2 Comparative population density.

	1981 population (000s)	Area (000 sq. km)	Population per sq. km.
Great Britain	54,743	229.9	238
West Germany	61,670	248.6	248
France	53,960	547.0	99

Because of its relatively low population density (Table 1.2) and few large cities, public transport is generally more difficult to provide in France than in West Germany or Britain. Added to these difficulties, there is less local public transport infrastructure in France (Table 1.3).

Table 1.3 Vehicles in use 31.12.83.

	Cars	Motorcycle/ moped	Buses
Great Britain	15,854,000	1,290,000	113,000
West Germany	24,688,843	2,949,849	71,000
France	20,600,000	5,150,000	62,000

Source: International Road Federation (1984).

Whereas, in Britain, many of those who begin to travel to work by car transfer from the bus, in France it is often from the motorcycle (Webster *et al.* 1985). In West Germany there has been heavy investment in both private and public transport, and railways are used as the main form of local public transport in the majority of large cities. In Britain and France, all cities rely heavily on buses for

Table 1.4 Local rail transport in Britain and France.

	Length of route (km)	Year first section opened
London	397	1863
Paris	198	1900
Tyne & Wear	55	1980
Lyon	15.4	1978
Marseille	15	1977
Lille	13.5	1983
Nantes	10.6	1985
Glasgow	10.4	1896

local public transport, although the bus is used much more for journeys to work in the UK than in France. Paris is the only city where there are more local public transport journeys by rail (*métro* and *RER*) than bus. In these two countries, substantial local transport by rail is restricted to a few cities (Table 1.4).

In West Germany, on the other hand, most large cities have a substantial *U-Bahn* or light rapid transit network (Table 1.5).

Table 1.5 Local rail transport in West Germany

	Length of route (km)	Year first section opened
Berlin (UB)	101.0	1902
Hamburg (UB)	92.7	1912
Hannover (LRT)	60.7	1975
Rhein-Ruhr (LRT)	53.9	1977
Köln (LRT)	41.6	1968
Munich (UB)	41.0	1972
Frankfurt (LRT)	40.3	1968
Stuttgart (LRT)	39.5	1962
Bonn-Ludwigshafen (LRT)	25.5	1975
Nürnberg (LRT)	15.0	1972
Wuppertal (LRT)	13.3	1901

One factor which has discouraged undergrounds or light rapid transit in British cities is the relatively low density of population. Even the 13 Inner London Boroughs have a density of only 7,790 persons per square kilometre (1981) and densities in many other cities are less. Three to four thousand persons per sq. km is common outside the inner city. Twelve thousand persons per sq. km is common in inner areas of French and West German cities but this is not always the case where rail transit has been built. Particularly in West Germany, suburban development is commonly concentrated around transport nodes or along routes. In British cities (and latterly in France too) much inter-War suburban housing was built uniformly at around ten or twelve houses per acre. Consequently there are few routes where a high capacity transport corridor could be planned with a large clientèle within easy walking distance.

So Britain, West Germany and France have been chosen as three large EEC industrial countries sharing many common issues but with some different solutions. The mid-1960s were a time when, in all three countries, the decentralizing effects of the car began to bite into public transport and city-centre planning, and when it was realized that solutions must be sought if city-centres were to remain in anything like their established forms. The solutions sought depended partly on the built inheritance – space to build urban roads – but perhaps even more on attitudes towards individual rights and collective benefits, individual versus public spending. This study has started out with with second-tier large provincial cities. These cities are facing some of the most difficult choices in public transport infrastructure for the future and some of the most acute problems of central area land use planning. These are used as a starting point for comparison with primate and smaller cities later in this book.

2 Central and local government

In this chapter we begin to look at the environment and the tools with which the planning of development and public transport are carried out. Here we are not so much concerned with what the policies are as with the contexts within which they are prepared, implemented, permitted, obstructed, prevented or promoted.

Requirements for planning

Some planning is carried out by local residents' groups or interest groups in all three countries. In West Germany it is fairly common for municipalities to assist them, even when in conflict with municipal proposals. Williams (1978) describes an example in Hannover where city officials presented both city proposals and those of a local group. Whilst few British local authorities would go as far as this, it is quite usual to give local groups access to information.

However, most planning of urban development and transport is carried out by local government, central government or other public bodies, and much of the remainder by private-sector consultants hired by central or local government. Town planning and public transport planning are largely functions of government, basically because of the need to coordinate and control public and private interests.

To carry out the planning of urban development and transport, there are certain requirements and characteristics of government which will allow or facilitate effective planning.

Local representation. Both town planning and public transport requirements are perhaps of greatest public concern at a very local scale where the neighbourhood effects of a development proposal or the need for a particular transport service are most clearly felt. As well as locally-elected politicians, there are many local residents' associations and interest groups. The dilemma in local representation is how to remain sensitive to local aspirations without allowing government to be paralyzed by the unreasonable veto of a sectional interest. In building any large-scale project such as a *métro* there will be some losers who, regardless of compensation, would prevent development if they were allowed to.

Local coordination. For public transport this means coordination for areas corresponding to local journeys to work, shopping, school, leisure and other journey purposes. For town planning it also means coordination of other requirements such as refuse disposal sites and other infrastructure. Whereas representation is compatible with small local government areas, requirements for coordination will pull in the opposite direction. Part of the task of coordination in transport will be regulation of competition to avoid duplication, to ensure that competition does not lead to inefficiency and to attempt to meet all journey requirements.

Decision-making on matters of more than local concern. As well as local coordination, on some matters such as nuclear power policy or airports there is a need for coordination at regional and national levels. This is partly because more than local interests are served, partly because such matters will be met very infrequently locally and therefore a national pool of expertise can be drawn upon. In Britain, applications for planning permission for matters of national significance are called in by the Secretary of State for the Environment or the Secretary of State for Scotland. In France, central government is represented locally by the *départements*, whilst in the Federal Republic of Germany central government has powers to become involved outside the planning legislation.

Arbitration in local disputes. Central government can act in cases of dispute between individual and local authority. In Britain, refusal of planning permission to carry out development or permission subject to allegedly unreasonable conditions carries with it the right of appeal to the Secretary of State for reconsideration of the case on a matter of policy. Central government and the courts have also acted as a second chamber on occasions such as the renowned case of the Greater London Council and the London Borough of Bromley on the matter of reduced fares.

Coordination of legislation. Lack of coordination may well lead to competition, with each municipality bidding to offer more favourable terms to developer or transport operator. This would obviously weaken their control and ability to meet other objectives such as recoupment of betterment. Coordination is generally a function of central government in Britain and France, but in West Germany the *Länder* (States) play a significant part as well.

Adequate research and professional skills for formulating policies. In all three countries, central government (and in West Germany also the *Länder*) often provides research skill for local government, partly to avoid repetition (many municipalities working simultaneously on similar projects), and partly because of the extent of staff resources needed. Where there are large numbers of small authorities as in France, providing professional skills for more routine jobs is also significant.

Powers of implementation including finance. Planning of all kinds needs powers of implementation to be effective. Where planning relies on controlling, coordinating and promoting the private sector, it can only be effective when there is demand from that sector. This has long been a serious weakness of town planning in some regions and it has become more widespread during the last decade or more. Otherwise, and particularly important in transport planning in which municipalities play a large role as operators as well as planners, an essential requirement of government and legislation is that adequate powers to raise finance should be available.

Evening out of local inabilities to raise finance. It is widely acknowledged that some regions are unattractive, perhaps due to outmoded infrastructure, unattractive environment or special locations such as nearness to the East German Border. In most Western countries, including all those examined here, it is accepted that without government intervention the differences in prosperity between regions would be intolerable.

Promotion and regulation of the private sector. In all capitalist and mixed economies, as are the three studied here, the externalities of the free market would be such that government intervention is necessary to prevent private profit resulting in unacceptable public costs. Implementation of much urban development and transport services relies heavily on private capital in all three countries but all acknowledge that it must be done in such a way as to avoid unacceptable public costs. For example, without regulation, speculation in land following a decision on public investment might result in a municipality being unable to reap the full benefits of its own investment by being unable to purchase land at the new elevated price.

The nine requirements listed above are intended to cover what is needed of governmental and legislative systems for planning to be effective. They are criteria by which local government systems can be assessed. Unfortunately, it is not possible neatly to allocate each of these requirements as being the responsibility of central government or local government or the legislative system. Local representation, for example, is a concern partly of the structure of local government, partly of the legislative system and very much of local planning.

Central government

Great Britain

The Department of the Environment has had the responsibility in England for the whole range of functions affecting the built and natural environments – the planning of land, housing construction, historic towns and buildings, preservation of coast and countryside, control of noise, air and water pollution, local government, regional policy, and at first, transport, but this is now a separate Department. In Wales, the Welsh Office has similar functions to the Departments of the Environment and Transport in England.

In Scotland, town planning and environmental responsibilities have come under the Scottish Development Department since 1962. In 1973 the Scottish Economic Planning Department was formed to take over responsibility for economic and industrial development including support to industry, new towns and regional development of Scotland within the UK and the EEC.

Central Government in the forms of the Secretaries of State and their Departments act both as policy-makers and as delegated legislators. Within limits set by Parliament, the Secretaries of State can make regulations such as the Use Classes Order (which defines industrial and other land use classes within which change can take place without the need for planning permission) and the General Development Order which contains a list of development for which planning permission is not needed.

A wide range of local authority proposals need approval by the appropriate Secretary of State. For example, structure plans need approval. If a local authority fails to make a plan, the Secretary of State can intervene to cause one to be prepared. The Local Government, Planning and Land Act 1980 allows the Secretary of State to direct an authority to make an assessment of land suitable for residential development. He or she can give directions about consultations the way they are to be made, and can require the publication of the results of these consultations, including public transport.

The Secretary of State can modify or revoke a local authority decision on an application for planning permission if the applicant appeals against the decision, and can 'call in' any application regarded as being sufficiently important – perhaps a new section of motorway or a new transport interchange. The decision of the Secretary of State on policy matters is final. Appeal can be made only on a point of law. However, the appeal procedure on policy or point of law is slow, a decision often taking more than a year. Developers may be reluctant to wait so long and so the local authority has a lot of power to modify planning applications even if the developer is convinced that the decision on appeal would be favourable.

The Department of the Environment, Welsh Office and Scottish Development Department also have important policy-making functions expressed in circulars, memoranda or bulletins, or implicit in their decisions. However, local application of policies lies with local authorities. There are many instances of local government failing to implement a central government policy with which they are not in agreement. For example, housing action areas, general improvement areas and other schemes for housing treatment have been implemented to very different degrees depending as much on the initiative of local government as on local housing conditions. In many cases where a local authority chooses not to implement a central government policy it would be impractical for central government to act. On the other hand, there are examples of polices where central government has dug its heels in. The sale of council houses to tenants has had high priority since the Conservative government came into power in 1979 and has been carried out against the wishes of many local authorities.

Central Government can also appoint *ad hoc* agencies for a specific task. New town development corporations and urban development corporations for inner city areas are examples of direct central government intervention to take over local government functions (Cullingworth 1982).

France

Like Britain, town planning and transport are the responsibilities of separate ministries, the *Ministère de l'Urbanisme et du Logement* (Ministry of Town Planning and Housing) and the *Ministère des Transports*. The French system is still a very centralized one whereby central government is involved in far more local detail than in Britain or West Germany. Central government in France is involved as a matter of normal practice, whereas in Britain it is involved only as an arbiter in disputes or in other exceptional circumstances. French central government involvement is partly via civil servants in the 95 *départements*, administrative areas each headed by a *préfet*.

The *préfet* is the local representative of central government and has authority over all central government sub-regional services.

Also important in plan preparation is the *Direction Départementale d'Equipement*, a group of officials in which civil engineers are prominent. They are responsible to the *préfet* rather than to the elected council of the *département*. Recent legislation on decentralization reduces the dominance of the *Direction Départementale d'Equipement* in local plans (*plans d'occupation des sols* and *zones d'aménagement concerté*) and also in development control. More influence has been given to the *communes* in both these fields. The 6,400 *communes* which had an approved *plan d'occupation des sols* on 1st October 1983 (the date on which the relevant part of the new legislation took effect) will particularly have more influence in granting permission for construction. In these *communes* the mayor can take such decisions, whilst in other *communes* the former procedures still have effect – the mayor delivers the construction permit in the name of the State on instructions from the *Direction Départementale d'Equipement*. Political actions of civil servants are explained in Mény (1982) and a summary of recent legislation is contained in *Cahiers Français* (1985b).

Thus French central government has been much more involved in plan preparation than the British government, and it remains to be seen how much this has been reduced. Until October 1983 when responsibility was transferred to the *communes, schémas directeurs d'infrastructure* were prepared by the State in consultation with the regions and local authorities for the long-term planning of infrastructure extensions and modernization. Central and local government have often been in conflict over objectives, and, as Chapuy (1984) suggests, this might be one reason for the large number of *ad hoc* agencies and commissions at all levels.

The *Delegation à l' Aménagement du Territoire et l'Action Régionale (DATAR)* was established in 1963 to decentralize from Paris and to develop tertiary industries in eight provincial towns *(Métropoles d'équilibre:* balancing cities) – Lyon/St Etienne, Marseille/Aix, Lille/Roubaix/Tourcoing, Nancy, Toulouse, Metz, Nantes/St Nazaire and Bordeaux (Riffault 1981). These cities were chosen to compete with Paris in attracting industries, to stimulate economic growth to revitalize their regions and to develop service and decision centres which would not rely on Paris. *DATAR* represents several ministries and is responsible to the Prime Minister. It is responsible for coordinating planning agencies to produce non-statutory sub-regional plans. It has distributed funds to public and private sectors for management and decentralization and some funds concerned with regional development premiums. In 1966 it gained charge of interpreting the national budget in regional terms.

Attached to *DATAR* are several types of regional organization including regional research institutes, *OREAM (Organisation d'Etudes d'Aménagement d'Aires Métropolitaines)*. These were set up in 1966 in each of the six conurbations of Lille/Dunkerque, Rouen/Le Havre, Nantes/St Nazaire, Lyon/St Etienne, Marseille/Aix and Nancy/Metz.

Although France is still highly centralized, President Mitterrand and the socialist majority in the First Chamber of Parliament elected in 1981 were engaged in legislation to decentralize. The decentralization legislation seems likely to extend the powers and responsibilities of *communes, départements* and regions in town planning and transport.

The Federal Republic of Germany

In contrast to the situation in France, the Government of the Federal Republic is very much decentralized, more so than that of Great Britain. Most of the executive powers lie with the eleven *Länder* (states) and the 8,506 *Gemeinden* (municipalities), and significant parts of the legislative powers are also with the *Länder*.

Whilst the Federal Government has exclusive powers to make legislation relating to foreign affairs, citizenship, freedom of movement, currency, customs, Federal Railways and airways, ports and telecommunications, for example, civil and criminal law and economic legislation is shared with the *Länder*.

In 1968 there was a constitutional reform which allowed the Federal Government to fund the improvement of regional structure, the improvement of farming structure, new universities, local traffic improvements and reconstruction of hospitals.

Land officials administer some of the laws passed by the *Bund* (Federal Government). The *Bund* has significant administrative capability only in a few fields – customs, the postal service, the armed forces and waterways. Under the Basic Law of 1949 which acts as a constitution, the Federal Government has only those powers specifically allocated to it. Any residual powers go to the *Länder*. Whilst the *Bundestag* (Federal Parliament) is elected directly for four years, the Second Chamber or *Bundesrat* is made up of representatives of the *Länder*. The *Bundesrat* has veto power in some fields of legislation which gives the *Länder* some power over Federal legislation.

The *Bundesministerium für Raumordnung, Bauwesen und Städtebau* (Federal Ministry for Regional Planning, Building and Urban Development) is responsible for the monitoring of Federal and regional development objectives and policies in order to achieve balanced development in the country as a whole. The Federal Law of 8th April 1965 states that equal conditions must be created everywhere, but as Hall (1982) remarked, the *Länder* are largely left to carry it out.

Apart from the legislative function of the Federal Government, much of the national planning has been concerned with the more peripheral rural districts, such as the border with East Germany, Emsland on the North Sea Coast and parts of Schleswig-Holstein, Lower Saxony, North Bavaria and the Eifel. Following Central Place Theory, development centres have been selected mainly in areas needing directed development and near the borders. Financial assistance has been provided by the Federal Government but is administered by the *Länder*.

State and local government

The Federal Republic of Germany

As Table 2.1 shows, the eleven *Länder* vary much in size and population. Hamburg, Bremen and West Berlin are city states. The others vary a lot in level of urbanization and industrialization. The boundaries do not in all cases correspond to the needs for addressing spatial problems. Some of the Hamburg suburbs are in Schleswig-Holstein and Lower Saxony. Emsland is a region with strong connections with the Ruhr (Nordrhein Westfalen) but lies in Lower Saxony.

All eleven *Länder* except Bavaria have a single chamber parliament *(Landtag)*. Each has its own state development programme. They plan the local road systems and large projects such as airports. Apart from trunk roads and railways, the Federal Government has little direct influence on town planning. It legislates and defines the procedures for planning but not its execution. The *Länder* have power to legislate and carry out legal provisions for state and regional planning, local government law, and some aspects of environmental protection, and they share education with the Federal Government. They have supervisory powers over the *Gemeinden* (municipalities) and approve local plans.

According to the Basic Law, the *Länder* must allow local government the right to run its own affairs. *Gemeinden* are responsible for the construction and maintenance of local roads, public transport, water and energy supply, housing, construction, building and maintenance of primary and secondary schools, sports facilities, hospitals, theatres and museums. In urban areas, *Gemeinden* carry out Federal and state laws locally. They have the power to regulate building activities by local bylaws, zoning and building plans.

Table 2.1 Populations, areas and local authorities within West German Länder.

Land	Sq. km	Population (thousands) 31.12.66	31.12.86	Kreisefrei Städte & Landkreise	Gemeinden
West Berlin	480.14	2,185	1,879	1	1
Hamburg	754.70	1,847	1,571	1	1
Bremen	404.23	750	654	2	2
Schleswig-Holstein	15,727.15	2,473	2,613	15	1,131
Lower Saxony	47,438.18	6,697	7,196	47	1,031
North Rhine Westphalia	34,067.88	16,835	16,677	54	396
Hesse	21,114.17	5,240	5,544	26	427
Rhineland-Palatinate	19,847.52	3,613	3,611	36	2,303
Baden-Württemburg	35,751.39	8,534	9,327	44	1,111
Bavaria	70,552.93	10,217	11,026	96	2,051
Saarland	2,569.34	1,132	1,042	6	52
Federal Republic	248,707.63	59,793	61,140	328	8,506

Source: Statistisches Bundesamt *Statistisches Jahrbuch für die Bundes-republik Deutschland 1987* Wiesbaden

There is still a large number of small *Gemeinden:* 6,389 have a population of less than 5,000. Many tasks are beyond the resources of the smaller *Gemeinden* and are carried out by the *Kreis*, the next higher level of local government. Otherwise the *Kreise* are responsible for matters covering a group of *Gemeinden* in common, for example, road making and local railways.

In West Germany, taxes are paid to central government and to the *Länder. Gemeinden* do, however, have considerable autonomy over certain resources and receive contributions from both *Bund* and *Länder. Gemeinden* as a whole receive 11% of total taxes (Bennett 1983). The *Bund* can only give grants to *Gemeinden* for investments, not current expenditure, and they must be specified grants in certain categories. *Länder* are obliged to provide grants to local government out of a percentage of their tax receipts, but the *Länder* decide this percentage and how precisely earmarked the grants are to be. All current revenue grants are distributed by the *Länder*.

The Federal Government controls about 50% of the total tax revenue. The other 50% is divided between the *Länder* and the *Gemeinden. Gemeinden* receive all the land tax, 60% of the manufacturing tax and 14% of the income tax. This revenue is, however, in many cases insufficient, and they frequently receive grants from the *Länder*. Such grants are mostly given only for defined purposes such as urban renewal and subsidies tend to favour regional rather than local interests.

Metropolitan transport authorities or *Verkehrsverbünde* were created for Hamburg (1965), Hannover (1970), Munich (1972), Frankfurt (1974), Stuttgart (1978) and Rhein-Ruhr (1980) to unify fares between enterprises (over 20 of them in the case of Rhein-Ruhr), to coordinate timetables and eliminate duplicated services. There are proposals to introduce further metropolitan transport authorities in Köln-Bonn, Rhein-Neckar and Nürnberg, but there is opposition from some municipalities fearing spread of debt.

There are also 45 *Tarifgemeinschaften*. In the Federal Republic, therefore, most public transport is in the public sector, and this accounts for 63% of the number of passenger kilometres travelled. The public sector is organized in various constitutional forms. Some are municipal departments, others are companies in which the municipality has a controlling share. Transport undertakings are organized into a national association, the *Verband Öffentlichen Verkehrsbetriebe* (Association of Public Transport Enterprises), which includes 168 public transport firms, the commuter lines of the *Deutsche Bundesbahn* (German National Railways), the Federal Post Office bus service and the rail and bus lines of 66 non-Federal public railways.

There are 4,900 privately owned transport firms, mostly small bus companies which are more important in transport between settlements than within. They are organized into the *Verband des Deutschen Personenverkehrsgewerbs* (National Association of German Passenger Transport Companies). Whilst private firms have only 21.8% of the total number of employees in transport, they generate 31.2% of the revenue. This is mainly because the law requires public

Table 2.2 Comparative public transport usage in West Germany.

	Million passenger kilometres (short distance only) 1977
Deutsche Bundesbahn	14,100.0
Non-Federal railways	517.0
Bus services of non-Federal railways	2,045.0
Bus services of the *deutsche Bundesbahn*	6,790.0
German Federal Postal Administration buses	582.0
Local transport enterprises	24,388.0
Private enterprises	31,394.0

Source: *Bundesminister für Verkehr* (1980).

undertakers to operate regular schedules whilst the private sector can operate charter buses with a much greater loading ratio and which produce more than half the revenue of the bus sector (Dunn 1981).

The Federal Government has rights of supervision, approval of new construction, economic planning, closing of lines, approval of time-tables and tariffs of the *Deutsche Bundesbahn* and the German Federal Postal Administration bus services.

The *Länder* approve the routes and timetables of all sectors except the *Deutsche Bundesbahn* and the tariffs of the non-Federal railways (including their bus services), local transport enterprises and private enterprises. *Gemeinden* have right of supervision of non-Federal railways and local transport enterprises where they are owners.

France

Below the level of national government there are 22 *régions*, 95 *départements* and 36,433 *communes*. Between *commune* and *départe-ment* there are also *arrondissements* and *cantons*, but neither are important in planning terms and the *arrondissements* do not have councils now.

Communautés urbaines were created in 1966 for Lille, Lyon, Strasbourg and Bordeaux to coordinate administratively a large number of *communes* and to give cohesion in financial, planning and transport services. They do not replace the legal powers of the *communes*. For planning there are also 28 *agences d'urbanisme*, or advisory planning agencies, for groups of *communes*, financed by central and local funds and responsible for preparing planning and policy studies. In new towns there are nine *établissements publics d'aménagement*, and there is a special planning agency in Paris (*Atelier Parisienne d'Urbanisme*).

The *départements* and *régions* (which are groupings of *départements*) are really extensions of central government, headed by a *préfet* (regional *préfets* were abolished in 1981), a civil servant appointed by the Minister of the Interior. *Départements* are approximately equal in area, defined so that officials can travel to any part of their territory in a day by stagecoach.

Under the *Loi d'orientation des transports intérieurs* (1982), the *région* has retained three important tasks: the preparation, with central government, of the *Schéma Régional des Transports*, the systematic development of public transport usage, in connection with *SNCF*, and the agreement of Contracts with the State for transport development. The Decentralization Law of 1982 gave the regions power to promote economic, social, cultural and scientific develop-ment in the planning of their territory. Since then they have had power to carry out their own services and to recruit adequate staff. The President of the Regional Council has taken over many of the powers of the regional *préfet*.

In the *région* Provence-Alpes-Côte d'Azur, for example, the *région* gives architectural and planning assistance to *communes* of less than 5,000 population and subsidizes the cost of preparing a *plan d'occupation des sols*. It also takes part in preliminary planning studies on greenfield sites and renovation. The *région* constructs around 8,000 dwellings per year (Conseil Regional Provence-Alpes-Côte d'Azur 1984 and 1985).

The council of the *commune* is elected for six years on a two-ballot, single-party list system, i.e. the party list which receives either an absolute majority or a relative majority on the second ballot receives all the seats. It is headed by a mayor who is both political leader and head of technical services. The mayor is appointed by the council from amongst its members and normally remains for the period of the council.

Many of the *communes* are far too small to employ adequate professional staff to carry out their planning, transport and other duties and they rely very much on the *département* or *région*. Although the responsibilities of the *communes* are similar legally, they depend in practice on local political will and size and consequently resources and technical expertise. In larger cities, the distribution of power between central and local government is fairly equal but in small *communes* the local councillors are hardly more than representatives of local interest to central government. Even in the larger towns, the power of *communes* is limited because central government has most of the technical expertise (d'Arcy and Jobert 1975).

Communes have recently gained more independence if they have the expertise and an approved *plan d'occupation des sols* or other plan *(Loi relative à la répartition des competences entre les communes, les départements, les régions et l'Etat 1983;* Act on the distribution of responsibilities between *communes, départements, régions* and the State). They may create joint authorities for services such as planning, transport, housing, sewerage and refuse disposal. They receive grants on a percentage basis for specific purposes such as these and also the revenue from a tax on wages.

It may seem that the *communes,* created in 1789, are today a particular anachronism. However, autonomy at the level of the *commune* is still anchored in the French sentiment. Some of the discussion on local government at a conference on *l'Urbanisme-Déplacements-Transports* (Town Planning-Journeys-Transport) organized by the *Fédération Nationale des Agences d'Urbanisme* (National Federation of Town Planning Authorities) at Lyon in 1981 was revealing on recent attitudes. It was agreed that government at the level of the conurbation is necessary but the majority of local politicians did not wish to see an intermediate level of government. The conference favoured *syndicats d'études* (joint boards on particular subjects) based on areas with compatible interests and common problems. These would need funds which would be used for projects involving more than one *commune*, and which would resolve the problems of different levels of funding and between planning and transport. The conference also called for more responsiveness from national organizations, such as the *SNCF*, in planning and transport. Inter-communal cooperation must be voluntary, flexible and practical, and must evolve from what already exists.

Transport authorities
In Paris (Chapter 9), Lyon, Marseille, Lille and Nantes there are conurbation-wide transport authorities with clearly-defined functions. We will look briefly at those in Lyon and Marseille as examples. The *Syndicat des Transports en Commun de la Région Lyonnaise (STCRL)* is the owner of the network and is responsible for its operation and development within the *Communauté Urbaine*. It is the policy making body, and has representatives of the *Département du Rhône* (4) and the *Communauté Urbaine* (4). It fixes the level of *versement transport*. It also has two executive companies: the *Société d'Economie Mixte du Métropolitaine de l'Agglomération Lyonnaise (SEMALY)* AND THE *Société Lyonnaise des Transports en Commun (STCL)*. *SEMALY* carries out research for construction of the network and for equipment. Research includes impact studies and research on transport needs and economics for example. *STCL* operates public transport.

In Marseille, the distinction between policy-making and operation is not quite so clear. The *Régie des Transports de Marseille (RTM)* operates all public transport. At its head, the Council comprises four representatives from the City Council, one *conseiller général*, two members nominated by the *Commissaire de la République*, one member of the Chamber of Commerce and two staff representatives. The *Société du Métro de Marseille (SMM)* is responsible for the construction of the *métro*.

The *Office de Coordination des Transports, de la Circulation et du Stationnement de Marseille (OCOTRAM)* was formed in 1901. Its aim is to promote overall policies for urban journeys based on priority for public transport. It is presided over by the Mayor and has representatives of the City of Marseille *(Agence d'Urbanisme, Direction Générale des Services Techniques), SNCF, SMM, RTM and Société Marseille Parc Auto (MPA)*. It has followed policies based on priority for public transport, such as provision of busways, priority at traffic lights and reorganization of the network. As part of the policy to control circulation and parking, it aims for 1,000 park and ride places at *métro* and tramway stations, 5,000 places for cars on the periphery of the city centre and 1,000 places for cycles and motorcycles.

Great Britain

The Greater London Council was formed in 1965 covering an area of 1,579 square kilometres and at the time of the 1981 Census had a population of 6,713,165. Within the area formerly covered by the GLC there are 32 London Boroughs and the City of London. The GLC was responsible for strategic planning, main highways and traffic control, overspill housing and ambulance and fire services. They prepared the strategic development plan or structure plan for the whole of their area setting out policies on transport, housing, employment and open spaces, for example. The London Boroughs prepared their own local plans within this framework. Central Government took away from the GLC the control of London Transport on 29th June 1984 and re-named it London Regional Transport. The GLC received from the Boroughs applications for development of strategic significance, such as transport terminals.

Under the Local Government Act 1972 a two-tier structure was set up, with six metropolitan counties in the conurbations containing 36 metropolitan districts and 39 other counties with 296 districts.

The metropolitan counties were abolished along with the GLC on 1st April 1986. Some of their powers (planning, highways, traffic management and waste disposal) were transferred to the districts. Others (fire service, police and public transport) became the responsibility of joint boards made up of district council representatives (and magistrates in the case of the police). The Inner London Education Authority remains in existence and became a directly elected body.

Of the 39 other counties, 22 came within the population range 400,000 to 700,000, nine in the range 800,000 to one million, and four (Hampshire, Kent, Essex and Lancashire) in the range 1.3 million to 1.5 million, in 1981. Counties prepare structure plans and agree a scheme for local plan preparation with the districts and are involved in major applications for planning permission and development control issues. They are transport authorities but none operate public transport services themselves. Blowers (1976) points out some of the inadequacies of their powers to meet the responsibilities they have. They have no power over nationalized operators such as British Rail and the National Bus Company, or over private companies. The Traffic Commissioners had independent powers of licensing and control of public transport until the Transport Acts of 1980 and 1985. Car-parking provision, including the granting of planning permission for it, is partly a district responsibility.

In functions other than planning and transport, the metropolitan districts have more obligations than the other districts. They are, for example, responsible for social services, education and libraries.

In addition to the districts, there are 10,950 parish councils in England which have the right to be consulted on applications for planning permission.

In Wales there are eight counties with 37 district councils and no metropolitan counties. The functions of county and district are very similar to those in England. There are no parish councils but there are community councils which have the right to be consulted on applications for planning permission.

Scotland has nine regions with 53 district councils. The regions are responsible for highways, public transport, education, regional planning, social services and water (which is the responsibility of nine water authorities in England and in Wales, the Welsh Water Authority). Districts have local planning duties, housing and a range of other functions.

Both counties and districts are planning authorities. In the three sparsely-populated regions of the Highlands, Dumfries and Galloway and the Borders, only the region has planning duties. In addition to the nine regions there are three Highland areas where there are no district councils.

There are provisions for community councils where there is a demand for them, under the supervision of the district councils or island authorities.

As well as the two tiers of local government, there are new town development corporations for the designing, planning and implementation of a new town. The development corporation consists of a board appointed by the Secretary of State and which employs technical staff. The board is not elected and is disbanded when the new town is well on the way to completion.

The Local Government, Planning and Land Act 1980 provides for similar organizations, urban development corporations, to direct renewal in run-down inner city districts. Two were declared almost immediately and five more recently. In a similar way in Scotland, the Scottish Development Agency has intervened in the renewal of the eastern part of Glasgow, the Glasgow Eastern Area Renewal Project, set up in 1977.

Passenger transport authorities

In England and Wales the counties and in Scotland the Regional and Island authorities are the passenger transport authorities. They are political bodies made up of representatives of local councils. The Transport Act 1968 set up passenger transport authorities for the West Midlands, Merseyside, SELNEC (South-East Lancashire North-East Cheshire) and Tyneside. To these four passenger transport authorities were added two further ones for West and South Yorkshire under the Local Government Act 1972. The Local Government (Scotland) Act 1973 established the Strathclyde Regional Council which became responsible for the Greater Glasgow Public Transport Authority (created in 1972).

The passenger transport executives in the former metropolitan counties and the Strathclyde region of Scotland are professional officers responsible for the planning, management and operation of public transport. Local public transport is also provided by British Rail, municipalities (district councils in England and Wales, regions in Scotland) and independents. Nationalized groups (the National Bus Company and the Scottish Bus Group) cover mostly longer distance routes. The usage of all sectors except private operators has declined in recent years.

Independence of local government

Compared with West Germany and some other countries, British local government does not have independence except by the will of central government. Recent changes are reducing what independence it has. An aim of the Rates Act 1984 was to reduce the possibility of local authorities financing increased spending by increasing the local rates. The Secretary of State for Scotland has had comparable powers since 1981. The Department of the Environment prepares an estimate of what central government thinks each local authority should spend, taking into account social and economic conditions. This Grant Related Expenditure Assessment determines the amount of grant from central government (the Rate Support Grant) to each local authority. If a local authority spends more than the Grant Related Expenditure Assessment it loses a proportional amount of its Rate Support Grant. The Secretary of State also has power to impose limits on the amount of rate which may be levied by local government and therefore the expenditure of each local authority. For the financial year 1985/86, eighteen local authorities were identified as being substantial overspenders and their expenditure was controlled, i.e. 'rate capped' (incidentally, sixteen of them were the opposite political colour to central government).

Town planning in the UK is very largely the duty of central and local government, and so shares, with other disciplines, many governmental powers and constraints. In many local authorities,

officials have very little freedom to exert any initiative or professional skill outside the confines of their office. In many cases, it is
not within their interest to do so. It would be unthinkable in many
authorities for an offical other than the chief officer to make a
recommendation in his or her own name. In some authorities, officials
are not allowed to sign a letter in their own name but are expected to
forge the chief officer's signature, or to ask the section leader to forge
it. As well as misleading the public, this seems symptomatic of a
rather archaic, inflexible and mechanical attitude towards local
government which often permeates planning even more noticeably
than some other areas of work. Williams (1978), for example, quotes
examples of West German multi-disciplinary teams composed of
members of various local authority departments who are able to
recommend policies in their own names rather than those of
departmental heads.

Eversley (1974) refers to the relatively low status and esteem in
which British local government officials are held, indicative of public
attitudes towards local government. The West German system is
more positively geared towards action than the British system, which
is fundamentally set up for stability. The more frequent changes in
political leadership in Britain aids this. British law encourages
officials and local politicians to keep performing their duties rather
than to use their initiative. Council members are personally
responsible to pay the cost of any enterprise which is not strictly
within their statutory powers.

Conclusions

The lowest level of government consists of 386 districts in Great
Britain, 8,506 *Gemeinden* in the Federal Republic of Germany and
36,433 *communes* in France. The small size of many *Gemeinden* and
communes means that they are weak and do not necessarily represent
local interests better than the much stronger district councils in
Britain. Certainly many *communes* are weak and central government
control over plan implementation strong. In Britain the districts are
made up of wards for which members of the district councils are
elected. Although wards have no powers, they do provide a means of
local representation to the district council.

The British structure of local government might suggest this to be
the most effective of the three. However, much depends on the way in
which it is operated. One of the limitations of British local
government is the public apathy towards it compared with West
Germany and France. The turnout at local elections is usually low in
Britain. Most of the memorable cases of public involvement are about

preventing development rather than promoting it. A protest about a new transport system is more likely to attract support than a campaign to build one. Even the abolition of the GLC and the metropolitan counties seemed to attract little interest.

Land use planning is the responsibility of the lowest level of local government in all three countries and, in Britain, the counties as well. In neither Britain nor West Germany is there the detailed intervention of central government found in France. There is, however, control over implementation by financial means. Districts, counties, *Gemeinden* and *communes* all rely on central government (and *Länder* too) for funds, and thus their powers of implementation are limited.

3 Planning and public transport legislation

Town planning, often referred to simply as 'planning', is not easily defined precisely. Traditionally it has centred on buildings and other physical development, such as roads and the laying out of open space. Along with architects and others, planners have long been concerned with the physical environment – the positioning and appearance of buildings and spaces between them. During the last two decades or so, there has been an increased awareness of the need to consider the ways in which the social and economic environment affect the physical environment. Some of town planners' work has been primarily addressed towards social and economic issues such as unemployment, with physical effects and issues being only secondary and often quite minor. In West Germany and France, as well as Britain, geographers, sociologists and economists are employed in planning in large numbers as well as those from the traditional, physically-orientated professions, principally architecture and, especially in France, civil engineering.

Planning has been practised at all scales from the whole nation to the individual building. In Britain, planning has been mostly at the scale of the county to the neighbourhood or street, although in development control, which probably occupies a third or more of practising planners, most work is on the single building, alongside architects, surveyors and building control officers. Regional and national physical planning in England has never been well developed. Many *ad hoc* regional and sub-regional strategies have been prepared (reviews are contained in Cowling and Steeley 1973) and Lichfield, Kettle and Whitbread 1975). Nevertheless, these have never been a statutory duty of local government, as are structure plans.

There is no national physical plan. National physical planning is reflected in ministerial circulars and other advice and in decisions by the Secretary of State for the Environment and the Secretaries of

State for Wales and Scotland, on appeals against refusal of planning permission or against conditions of planning permissions. Some regional planning functions are carried out by the Department of the Environment regional offices, the Scottish Office and Welsh Office.

By contrast, regional planning has long been well developed in both France and West Germany. In West Germany there is a hierarchy of plans for national, state, regions of states and smaller scale *Bauleitpläne* (literally 'plans leading building'). In France too there is a National Plan, regional plans and several *ad hoc* bodies concerned with national and regional planning (Chapter 2) as well as plans more equivalent in scale to British structure plans and local plans.

All of these scales of planning will have an important effect on the city centre. This is very clear in Central Place hierarchies expressed in West German regional and state plans. So, too, the policies of decentralization from Paris, pursued over several decades and planned by *ad hoc* bodies, will clearly affect the development of city centres. All of these scales must therefore be considered in the planning of city centres of not only the largest cities but a long way down the Central Place hierarchy.

Aims and purposes of planning

Planning has developed as a government response to perceived problems and issues. The urban sprawl and ribbon development of the 1920s and 1930s was the main reason for the foundation of the modern British land use planning system in 1947. For a long time, the aims of planning below the scale of the region were largely environmental, although it has long been recognized that costs of building and perhaps more significant costs of servicing, including related land uses such as schools, can be greatly affected by choice of site, location and the pattern of development (Simpson 1983, Simpson and Purdy 1984). It took a long time to acknowledge the fallacy of the assumption that social problems will be resolved in the wake of environmental and physical improvements.

Planning can be defined as the steering of future development and activities to meet desired ends. Loosely defined, planning can be undertaken by any organization. In the sense usually implied in town planning, it is very largely an activity of central and local government. Then it may be viewed as an attempt to address the failure of the market to make all costs of development accountable to the developer (for example visual intrusion, traffic nuisances from a profitable development) and the failure of the market to achieve publicly desirable ends. Uncontrolled private enterprise has made private profits at public cost, hence the proliferation of public health, housing, planning and other government legislation since the mid-nineteenth century to internalize, prevent or achieve compensation for these public costs.

A fundamental principle for West German national planning is stated in the *Raumordnungsgesetz* as 'to provide in all districts, and for everyone, equality of opportunity'. French planning too has certainly moved towards this sentiment in the 1980s with the coming into force of legislation covering many aspects of decentralization. It is true, however, that for centuries France has been far more centralized on Paris than Germany was on Berlin or any other city, and that the recent measures have only just begun to redress the imbalance of power and activity. Whether planning has striven for equality of opportunity in Britain is debatable. Many critics of British planning, in particular those with far left wing political views, have accused planning of achieving or even aiming for quite the opposite: 'planning has done nothing more than prop up land values'. Certainly development control has been, amongst other functions, a way of acting as an 'umpire' between neighbours. It has prevented a neighbour's unneighbourly conduct – cramming another house in the back garden or opening a car repair business in the garage, for example. Planning has been associated with the deliberate increase of land values such as in the large office centre of La Défense on the western fringe of Central Paris and the many British town-centre shopping developments of the 1960s. Of course town planning has helped to realize individual profits. It has also helped realize communal profits and tax revenue, and has mostly, if not always, restrained private profit to meet at least some public costs. Much planning implements central government policies related to equality of opportunity. Industrial development grants are an example. Nevertheless, the German principle of equality is not one which many British planners would use to explain the purpose of their work. It happens to some extent, but is not a guiding principle.

A second fundamental purpose of planning expressed in the West German *Raumordnungsgesetz* is that overall spatial development should be that which best serves the free development of the individual in society. Natural opportunities as well as economic, social and cultural provisions should be considered. This can be interpreted in many ways. Rather like many structure plan policies, it makes more sense viewed in relation to a specific proposal for development than without. Certainly there are large areas of work of French and British planners which could be interpreted as being towards this end. The development of many recreational, cultural and educational facilities could be interpreted as being for individual development. However, in Britain, it is probably the more common view of planning to see it as a means of safeguarding the interests of society *against* the free will of the individual. A large part of development control work is about protecting the interests of neighbours, the neighbourhood and the wider area against the failure of the market to result in *all* external costs being borne by those who promote the physical development which causes them. In some respects the British allow much more individual freedom than the Germans and French. Subsidies for public transport of 50% or 60% are common in West Germany and France, whilst half of that would be high for Britain. The British certainly have more freedom to spend their money as they choose rather than to subsidize expensive railway networks like those of the larger German and French cities.

National planning

Both West Germany and France have published national plans; Britain has not. The Ninth National Plan for France 1984-88 includes objectives and policies which will guide town planning and transport planning. Even more importantly, it can be used by developers, municipalities and other organizations to muster support for their development intentions. Some of the policies are as follows.

1 Modernization of industry and reduction of dependence on energy.
2 Better integration between town planning and transport in urban areas by legislation or advice.
3 Implementation of means to secure the right to transport for all, set out in the *Loi d'orientation des transports intérieurs 1982*.
4 Improvement of urban public transport networks to compete with the private car. An important means of improvement will be integration between modes of public transport, including transport interchanges to allow for the continuity of journeys.

The *métros* in Marseille, Lyon and Lille clearly reflect these policies.
    A series of *Grundsätze* were published in 1962 to give advice on the spatial structure of the Federal Republic as a whole and these were forerunners of the *Raumordnungsgesetz*. Following on from the principles of the *Raumordnungsgesetz* a series of special areas were defined.

1 *Zuruckgebliebenen Gebiete* (less developed areas), defined on criteria such as income, population density, degree of industrialization, ability to raise local taxes and value of product per person employed.
2 *Zonenrandgebiet*, where economic development was affected or

threatened by nearness to the East German boundary (19% of the Federal Republic).

3 *Erholungsgebiete* (recreation areas). With the recognition of the increase in stress, income and holidays, areas have been defined to serve the local area or a wider catchment area, such as Aachen, Wiesbaden and Stuttgart-Bad Cannstatt.

The principles of the *Raumordnungsgesetz* were further developed in the *Bundesraumordnungsprogramme 1975* which laid down a model for the development of each *Land*, conformable to the development of West Germany as a whole, and set out issues, policies and priorities. More explicitly a spatial planning document than the French National Plans, it has been remarked by Kunzmann (1984) that day-to-day problems have come to take precedence in urban and regional planning and that the *Bundesraumordnungsplan* has become almost forgotten.

Britain has never had a national spatial plan in the form of a written document. Advice and policies equivalent to those of the French and German national plans are scattered amongst many government circulars, bulletins and other documents. Much is also implicit in ministerial decisions, in connection with appeals against refusal of planning permission, for example. It would be hard to demonstrate that this dispersed nature of central government policies and advice has been detrimental to any aspect of planning or that the absence of a national plan has resulted in less national planning.

Regional planning

In West Germany there is a clear hierarchy of plan forms from the whole of the Federal Republic down to the individual building. At the regional scale there are *Landesentwicklungspläne* and *Landesentwicklungsprogramme* (development plan and development programme for the *Land* or state) and *regionale Raumordnungspläne* (development plans for parts of the *Land*). These are prepared by the *Land* in consultation with the *Gemeinden*. They include social, cultural, economic, and ecological requirements and objectives for *Bauleitpläne (Flächennutzungspläne* and *Bebauungspläne)* and they identify major land use designations as well as policies.

Germany was associated with the pioneering work on Central Place Theory. (A useful review is contained in Szumeluk 1968.) It is not unexpected that the factors which made it a suitable study area earlier this century should cause the theory to be still applicable in regional planning. The Central Place hierarchy is the basis of many of these regional plans with clear implications for policy and the strategic planning of city centres. For example, the regional plan for the *Land* of Baden-Württemburg identifies a hierarchy of *Oberzentren* and *Mittelzentren* centred on Stuttgart. Transport and infrastructure corridors are to be improved to enable axial development, whilst sporadic development around villages is to be strictly controlled. As a means of improving services in the Black Forest and Upper Swabia, Villingen-Schwenningen is to be upgraded to one of the 14 *Oberzentren* (the highest category of Central Place below Stuttgart: Scott 1983).

Structure planning

Below the level of the region there is the structure plan in Britain, the *Flächennutzungsplan* in West Germany and the *Schéma Directeur d'Aménagement et d'Urbanisme* in France, all of which are concerned with the configuration of land uses, albeit with varying degrees of precision. All cover a substantial period: *Flächennutzungspläne* 7 to 12 years (von Klitzing 1974); structure plans 15 to 20 years (Department of the Environment 1979); *Schémas* up to 30 years (Ministère de l'Environnement et du Cadre de Vie, Service de l'Information 1979). Structure plans and *SDAUs* define the general pattern and relationships between land uses and communications but the structure plan in particular was originally intended not to be used to define precise land parcels. Both forms of plan are used to define the locations of future major urban development, major renovation works, open spaces to be protected, major new infrastructure, guidance for local plans and to set out municipal policies and priorities towards development. Introduced under the *Loi d'orientation foncière* (Land Guidelines Act) 1967, *SDAUs* generally cover smaller areas than structure plans. They must be prepared for *communes* of more than 10,000 people (about 7,000 out of 36,433). They may be for two or more neighbouring *communes*. By late 1980, 370 had been started covering about 26% of the area of France and 72% of the population, although progress had been slow and only 160 had been approved.

Structure plans were introduced under the Town and Country Planning Act 1968. They are mostly for whole counties in England and Wales, although some are for divisions of counties and urban areas. By late 1982 all county authorities in England and Wales had submitted structure plans to the Secretary of State and all except six had been approved. Structure plans have varied a lot in content. Some of them, such as those for West Yorkshire and Merseyside, have been used as a means of adding weight to social and economic policies with no direct bearing on land use. Many others have moved substantially away from the original intention as strategic planning documents.

In Scotland, as well as structure plans there are also regional reports. These are important in negotiations between the Regions and the Secretary of State for Scotland and they give direct guidance on planning policies to district authorities and developers.

The West German *Flächennutzungsplan*, prepared under the *Bundesbaugesetz* (Federal Building Act) 1960, also sets out land uses, generally for a whole *Gemeinde* or sometimes for a wider area. Being generally for smaller areas than structure plans, they set out land uses in a rather more precise way. Some are presented in diagrammatic form, but usually land areas are identifiable. Otherwise, the level of detail on land uses is comparable with that in a structure plan. Structure plans, and to a lesser extent *SDAUs* are, however, far more concerned with policies and with discussion of alternatives. Many *Flächennutzungspläne* are rather bald statements of land use proposals (as their name implies) with very little discussion of the urban processes within the study area which caused the factors affecting the proposals. They are more reminiscent of the British pre-1968 development plans than the structure plans which replaced them. Perhaps because in West Germany there is a well-defined statutory hierarchy of regional plans, the *Flächennutzungsplan* is less concerned with strategic issues than the structure plan and *SDAU*. It is significant that, along with *Bebauungspläne* (local plans), they are referred to as *Bauleitpläne* (literally building leading plans).

Flächennutzungspläne are statutory plans. *Rahmenplanungen* (sometimes referred to as *Strukturpläne*) and *Entwicklungsplanungen* (development plans) are non-statutory plans which like *Flachennutzungspläne*, act as guides for *Bebauungspläne* (local plans). *Rahmenplanungen* often cover quite small areas, parts of a *Gemeinde* and set out as much detail as might be contained within a district plan in Britain. Many *Entwicklungsplanungen* contain much of the discussion missing from the statutory *Flächennutzungspläne*.

Public involvement in structure plans

In Britain there have been statutory measures for publicity, allowing public comment and public discussion of structure plans since they were first introduced. The Structure and Local Plans Regulations lay down minimum publicity measures. After a structure plan has been submitted to the Secretary of State for the Environment an Examination in Public takes place. The Secretary of State appoints an inspector and panel to preside over this Examination and they report back to the Secretary of State, giving details of discussion and publicity measures. The inspector representing the Secretary of State decides which policy issues are to be discussed and which representations are to be heard, but those comments which are heard must be considered.

Until recently, there was no obligation to allow for any kind of public involvement in preparing an *SDAU* or even to make it known that one was being prepared. Recent decentralization legislation now makes it obligatory for a draft *SDAU*, and comments on it made by public authorities, to be published *(Loi relative à la répartition de compétences entre les communes, les départements, les régions et l'Etat:* Act on the distribution of responsibilities between *communes, départements, régions* and the State, 1983). It has been commented that at least until recently, *SDAUs* were often not inspired locally but were more or less imposed by the State, a sentiment reflected by M. Roger Quillot, the Minister of Town Planning and Housing, at a conference on Town Planning and Transport at Lyon in 1981 *(Fédération Nationale des Agences d'Urbanisme 1981)*.

Under the *Bundesbaugesetz 1960* (S 2-7), a *Gemeinde* must make provisions for publicity and discussion of all *Bauleitpläne (Flächen-nutzungspläne* and *Bebauungspläne)*. Plans must be available to the public for inspection for at least one month. If more than a hundred people comment on what is substantially the same subject, the consideration which has been given to these comments must be published. A final check on what consideration has been given to public comment lies with the *Regierungspräsidenten*, who has to be satisfied that justified comments have not been overlooked before approving a plan.

Local planning

In all three countries there are local plans which relate structure plan policies and proposals to precisely-defined areas of land, which provide a detailed framework for coordinating land use and development control, and which draw attention to detailed planning issues for discussion by public and politicians. In Britain, local plans are advisory and give no guarantee that development in accordance with them will get planning permission. In France and West Germany, the *plan d'occupation des sols* (literally land utilization plan) and *Bebauungsplan*, on the other hand, define the legal rights attaching to each parcel of land. Both French and West German local plans are more rigid, legal documents, less concerned with understanding issues and with promoting public discussion than are British local plans, which are far removed from the lawyers. Most British local plans are not even prepared under any legislation but are drawn up informally, more concerned to promote debate rather than lay down regulation.   In West Germany, the *Länder* can determine some of the detailed issues which a *Bebauungsplan* must address. It is common for them to set out a lot of architectural details, for example building lines, roof forms, building materials and plot ratios. These can be defined in British local plans, but it is more common for them to be left as a matter for discussion as a part of development control after applications for planning permission have been received. In *Bebauungspläne* housing can be regulated as detached, semi-detached or terraces of specified lengths. Land uses can be specified in categories described in the *Baunutzungsverordnung* (Ordnance on the use of land for buildings). The West German local plan system therefore allows for the setting-out of a lot of detail which in Britain is now usually left open for negotiation. The French *plan d'occupation des sols* also defines details such as building lines, regulations for access and parking requirements with more detailed instructions for developers than do local plans in Britain. On the other hand, West German and French local plans allow for less innovation and creativity from the private sector.

Most new road development is not regulated by the *Bebauungsplan*. For a new by-pass or tunnel, for example, it would be more usual to use *Planfeststellungsverfahren* (plan establishment) procedure. The procedure would be similar for, say, a new *U-Bahn* or an airport.

A *POS* must be prepared for all settlements of more than 10,000 people, for *communes* where there is large scale redevelopment or renovation (except where included in a *zone d'aménagement concerté)* areas of major environmental improvements, areas peripheral to national parks and listed tourist areas. Formal local plans in Britain, on the other hand, are prepared as a result of a Development Plan Scheme agreed between the county planning authority and the districts under the Local Government Act 1972. At the end of March 1982, only 230 had been completed in England (13 in London [Field 1983], and 217 in the rest of England [Bruton 1983], although a further 1,463 were being contemplated. By late 1980, 11,403 *POS* had been designated. Although comparisons are fraught with imprecision because of the differences in sizes of the study areas, it is clear that formal local plans in Britain cover a much smaller part of the country than do *POS* in France. They are not a routine task but a local response to a perceived need for a plan.

As well as designating land uses and infrastructure, the *POS* also defines certain legal rights attached to land: access rights, conditions attached to demolition and building lines and even conditions of transfer of land in certain areas of high landscape value.

British local plans are of three kinds:

1 Action area plans for areas where intensive change by development, redevelopment or improvement by public or private developers is envisaged.

2 District plans which typically cover larger areas such as the whole
of a small town or district of a city with a full range of local issues.

3 Subject plans, which are confined to one type of development such
as housing, industry or open space.

Neither French nor West German local plans have quite the same
explicit division, although *Rahmenplanungen* and *Entwicklungsplanungen* fulfil some of the functions of subject plans. The French
zones d'aménagement concerté (Loi d'orientation foncière 1967) often
cover areas which in Britain would probably most usually be declared
action areas although they are not necessarily prepared by the
communes. They can also be prepared by the *département*, State,
établissement public (public authority) or by a private developer.
When undertaken by a developer in the private sector, a formal
agreement for the *ZAC* has to be made with the *commune* or other
public authority. A *ZAC* is approved by the *préfet* except for those
involving renovation or building of at least 10,000 houses, where a
State subsidy is needed to finance infrastructure or where an
abnormal subsidy is requested by the *commune* or by the developer.
In those cases, *ZAC* approval must be by the *Conseil de Direction du
Fonds de Développement Economique et Social* advised by the *préfet*,
or, in the case of *ZACs* for industry, commerce or port uses, by the
minister responsible for town planning.

British local plans, on the other hand, are normally adopted by the
local council although they must allow for 'adequate publicity' and
give 'adequate opportunity' for members of the public to make
representations (Town and Country Planning Structure and Local
Plans Regulations). Inquiries may be held by inspectors of the
Department of the Environment, but these inspectors report to the
local council, not to the Secretary of State for the Environment. Local
plans in Britain are therefore much more of an 'in house' production
of the local authority than is the French *plan d'occupation des sols*.
This is perhaps to be expected when we remember the smallness of
many French *communes*.

In West Germany, the *Bundesbaugesetz 1960*, as well as providing
for *Flächennutzungspläne* and *Bebauungspläne*, also introduced
regulations for compulsory purchase, pre-emption and reallotment of
land and buildings and rules for the distribution of development costs
(Albers 1986). Under the *Städtebauförderungsgesetz* (Urban Development Act) 1971, *Sanierungsgebiete* (renewal areas) may be declared
for areas requiring special urban renewal measures, to correct
planning deficiencies or for newly developing areas. They are mostly

for inner-city areas, and unlike *Bebauungspläne*, they have fixed target dates for completion. The *Städtebauförderungsgesetz* 1971 also allows for many powers to secure planning and implementation of renewal programmes. It has no doubt helped to retain population in the city centre and has affected transport demand significantly. The Act lays down criteria for reconstruction and redevelopment, plot standards, measures for relocation of residents, places of employment and replacement of infrastructure and facilities such as open space, kindergartens and schools. Property owners may form associations or companies to promote renewal. Municipalities may delegate to housing trusts or associations functions such as land and property acquisition. Profits are limited and municipalities may form valuation panels to control prices (Rosner 1975). Municipalities may freeze land values and may take powers to control land transactions. Owners can be required to alter or improve their property in accordance with an approved plan. They may be required to pay betterment levies, but compensation is payable for adverse effects on properties.

Under the *Städtebauförderungsgesetz* a social plan must be prepared as part of a renewal programme. This should specify effects of renewal on relevant groups such as owners, landlords, the elderly and children. Generally, other forms of West German plan are not characterized by their discussion of social and economic issues, being more physically orientated than their British counterparts. Perhaps this explains why it has been deemed necessary to make a social plan an explicit part of legislation in this case.

The *Städtebauförderungsgesetz* also allows powers for planned expansion of existing settlements, new towns and land purchase. It allows for rehabilitation in conjunction with reform of agrarian structure in rural areas.

The powers in *Sanierungsgebiete* are therefore more comprehensive than in action areas under British planning legislation. Like the French *zone d'aménagement concerté*, *Sanierungsgebiete* confer on the municipality a range of positive powers for implementation, far more than the more passive *Bebauungsplan* and *plan d'occupation des sols*. Regarding urban areas, comparable powers exist in Britain but they are scattered under other legislation as well as the Town and Country Planning Act 1971. Many housing problems similar to those in *Sanierungsgebiete* have been addressed within General Improvement Areas and Housing Action Areas under the Housing Act 1974. The Public Health Act 1936, and in respect of betterment and land prices, the Development Land Tax Act 1976 and the Finance Act 1979 have also been used to address comparable problems.

Control of land prices and speculation is a fundamental concern of planning in all three countries. It is an issue particularly significant in relation to transport, because transport infrastructure can greatly affect land prices. Thus the opening of a metro will create increases, and perhaps decreases away from the stations. The concern of local and central government has been how to recoup some of these increases for the public purse, which in many cases has been responsible for a large part of the costs of construction, without being so tough on the private sector that development will be stifled.

As we have seen, the *Städtebauförderungsgesetz* contains several measures to address this issue. In France there are several types of local plan, each with complementary powers to address the problem of betterment. In a *zone d'aménagement concerté*, developers may be required to construct public goods such as schools, although this has recouped only a small part of the profits of the private sector (Mény 1982). Also in a *ZAC* land is exempt from the *taxe d'urbanisation*, a tax on underdeveloped land. Land owners can insist that the municipality or developer responsible for implementation should purchase the land.

ZACs were instituted under the *Loi d'orientation foncière 1967*. *Zones d'aménagement différé* were instituted in 1962 mainly for extensions to urban areas, new towns, tourist areas and prior to urban renewal. When a land owner decides to sell, the public authority or designated private sector developer has the right to purchase land at existing use price one year before declaration of the *ZAD*. This provision remains for 14 years after declaration.

The designation of a *périmetre sensible* can be used to control land prices around large scale development as along the coast of Provence-Côte d'Azur, for example. This procedure gives the *département* special acquisition rights, powers to levy taxes on division of land and construction and increased control over construction, demolition, road-making, water collection and tree felling. Most have been created near to the coast.

Whereas *ZADs* controlled land prices, mainly on the periphery of urban areas, *zones d'intervention foncière* were introduced in 1975 to extend their powers and to give municipalities the right to be informed of land and property transactions. They could acquire land and property to further planning and social policies including conservation.

Unlike West Germany and France, the betterment problem in Britain has not been addressed by special powers attached to local plans. Betterment legislation applies irrespective of whether a local plan has been prepared or not. Under the Development Land Tax Act 1976 and the Finance Act 1979 there is a levy of 60% on the estimated increase in land values as a result of granting planning permission. Under section 52 of the Town and Country Planning Act 1971 a local planning authority has the power to make agreements with private-sector developers. Local planning authorities have had a lot of freedom to determine the scope of these agreements. They have been used to make developers carry out what at the time appeared to be unprofitable development, such as housing in unpopular districts, as a condition of planning permission for a profitable development.

In all three countries, some of the proceeds of increase in land values is recouped by way of the general taxation system on company profits, personal income, death duties on estates, local taxes such as rates in the UK and in many other diverse ways.

Apart from district and action area plans, the third form of local plan in Britain is the subject plan. As their name implies, these cover policies and proposals only for a single, even if wide-ranging, topic such as housing, open space, industry or public transport, or they can be more restricted such as for a policy on the conversion of shops to building societies. Their weakness, of course, is that if a single topic is seen in isolation there is a danger of overlooking alternative and perhaps more beneficial uses. For example, it may seem perfectly viable to designate a site for housing in a residential subject plan, but had an open space subject plan been prepared, the site may have been shown to be even more beneficial for that use.

In France there is no formal equivalent of the subject plan, although as in Britain many topic papers and policies are prepared on an informal basis, not governed by legal requirements and specifications. In West Germany, *Entwicklungsplanungen* (development plans) usually cover the whole of a settlement for subjects such as commerce or population. Unlike subject plans, they have no statutory basis. The *Freiflächenentwicklungsplan* (open space development plan) *1984* for Frankfurt, for example, covers general principles and detailed land use allocations for all kinds of open space including agriculture, leisure, allotments, forests and ecological protection. *Entwicklungsplanungen* usually serve as a guide to *Bebauungspläne* and other *Bauleitpläne* and are therefore hardly local plans in the British sense. However, subject plans, although classified by the Department of the Environment as a form of local plan, in many cases cover study areas much larger than action areas or even district plans.

Transport legislation and plans

In all three countries the physical land use planning systems are able to control the construction of car parking and other transport infrastructure which will affect the number of journeys and mode of transport. Control of the number of private sector parking spaces and road construction has been an important, perhaps the main, form of traffic restraint in many large cities. Provision of attractive cycleways in many German cities has probably provided competition for public transport. Alongside the land use planning system in each of the three countries there operates separate legislation for the control, management and operation of public and private transport, for the subsidy of capital works and operating costs, the regulation of fares and therefore the effects on demand for journeys and the regulation of competition and the achievement of coordination of services.

In Britain, the Transport Act 1968 allowed for 75% grant from central government towards the cost of public transport infrastructure, the same as was already available for road construction. Under the Local Government, Planning and Land Act 1980, a national cash limit on local authority expenditure was to take effect from 1st April 1981. Local authorities were invited to prepare each year programmes for expenditure on up to five services – housing, education, personal social services, transport and 'others'. As long as it stays within the total allocated, a local authority is allowed to switch expenditure from one service to another. The main submission under transport is the Transport Policies and Programme. These annual plans have been submitted by local to central government since 1975. They include transport planning objectives, priorities and programmes for all modes of transport, pricing and expenditure proposals. They must be compatible with the structure plan, but they are prepared with much less publicity and consultation and there is no Examination in Public. Perhaps the main function of the TPP now is to make the local authority's bid for the Transport Supplementary Grant from central government. Since April 1985 this grant has been only for road improvements, not for general transport expenditure. Since then, TPPs have become a part of the process of bidding for finance for road construction, and public transport has had to rely on local taxes (rates). Central government has control over total local authority expenditure. For each local authority, a Grant Related Expenditure Assessment based on its needs is approved by central government. If this expenditure is exceeded by more than a fixed percentage, the local authority's Rate Support Grant is reduced, i.e. it is 'rate capped'.

A more direct control over local public transport fares followed from the Transport Act 1983. This Act imposed Protected Expenditure Limits in the six metropolitan counties and the Greater London Council (all abolished on 1st April 1986). This followed reductions in public transport fares by the GLC and several of the metropolitan counties in 1981, subsidized out of the rates. The London Borough of Bromley challenged the reduction of fares and rate subsidy. Finally the Law Lords decided that this subsidy was illegal and the fares were increased. A useful blow-by-blow account of the whole episode is contained in Wistrich 1983.

Annual Public Transport Plans were introduced under the Transport Act 1978, at first for the non-metropolitan counties only. However, as Rigby (1980) pointed out, finance for public transport continued to be the concern of transport policies and programmes and county councils have inadequate powers to put into practice the improvements put forward in PTPs.

Unlike the situation in Britain, in both France and West Germany there are earmarked taxes for public transport subsidy, *versement transport* and *Mineralölsteuer*. *Versement transport* was introduced under the Law of 11th July 1973 as a local tax in all towns of more than 100,000 population (extended to *communes* of 30,000 inhabitants since 1982) and is charged to firms employing ten or more people. The level varies from city to city (2% in Paris, 1.5% in Lyon, for example). The percentage is charged to the firms on the salaries of employees up to a ceiling (the *securité social*) determined nationally each year. For instance, in a year where the ceiling was 80,000 francs, for an employee earning 50,000 francs the firm would have to pay 1.5% of 50,000 francs (in Lyon, for example). For an employee earning 100,000 francs, the firm would be charged 1.5% of 80,000 francs.

The money from *versement transport* has been used to finance capital works including the new *métros* in Marseille, Lyon, Lille and Nantes. It has also been used to repay loans for investments or network extensions or other improvements to services and as compensation to operators for reduced fares for travel to work. In 1980, *versement transport* provided 3,134 million francs out of a total of 7,710 million francs subsidy in the Paris Region and 1,876 out of a total of 3,638 million francs in other cities (Hanappe 1983). Durand and Pêcheur (1985) point to the differences in services and usage of public transport in towns which have levied *versement transport* and those which have not. In 1983 levels of both service and usage were about twice as high in those towns which levied *versement transport*.

Articles 7 and 15 of the *Loi d'Orientation des Transports Intérieurs 1982* allows for contributions to *versement transport* by all who gain from public transport services. Fournie and Pêcheur (1985) argue that, since 1982, *versement transport* has not been used for its original purposes but has just been used to fill in deficits in transport authorities. They argue that the principle of employer contribution of *versement transport* should be extended to shops as they benefit from public transport bringing in their customers as do industrial firms. As the authors point out, there would need to be safeguards such as for small neighbourhood shops whose customers are mostly so local that they walk.

The *Mineralölsteuer* (mineral oil tax) in West Germany is 6 pfennigs per litre. Not all this tax is used for local transport. In 1980, 11% was used for this purpose, 51% went into general tax revenues, 28% was used for trunk roads and 11% for other transport subsidies. In 1982, *Mineralölsteuer* raised 2,290 million DM out of a total subsidy of 2,574 million DM (Bundesminister für Verkehr 1984). This was about 20% more than *versement transport* in France, including inflation between 1980 and 1982. The proceeds of *Mineralölsteuer* are used equally for both local road construction and local public transport.

Under the *Gemeindeverkehrsfinanzierungsgesetz* (Municipal Transport Finance Act) *1975* projects can be financed by up to 60% by the Federal Government (75% along the border with East Germany). In 1982, the Federal Government provided 58%, the *Länder* 17.5% and the *Gemeinden* 24% of the money for capital projects. 88.7% of the expenditure was within the catchment areas of the eleven largest cities.

Between 1967 and 1981, 7,600 million DM were spent on *S-Bahnen* (67% from the Federal Government, 25% from the *Länder* and 8% from the *Gemeinden* [Girnau 1983]). In recent years, *Deutsche Bundesbahn* has had over four times the subsidy of British Rail, and moreover planning has been allowed on a more secure basis. Five-year plans of investment are approved by the Federal Government, whereas each project for British Rail has to be approved separately (Abbott 1985).

Funds for local road construction are distributed amongst the *Länder* in proportion to the number of vehicles registered there. The *Länder* draw up programmes for public transport investment and local road construction, but projects over 5 million DM need Federal Government approval. The *Bundesverkehrswegeplan* (Federal Transport Route Plan) *1980* envisages a shift from trunk road investment to the railways and local transport up to 1990, continuing the concern for public transport shown in earlier *Bundesverkehrswegepläne*.

Parking charges are under the control of the *Gemeinden* within the framework of the *Strassenverkehrsgesetz* (Urban Traffic Act) *1980*. The *Landesregierung* (State government) is able to set upper limits for charges (Muthesius 1982). The aim in setting rates is usually to promote the turnover of parking spaces to allow the greatest number of users.

Land use planning control over transport infrastructure and parking and legislation on transport finance will clearly have an effect on the development of commerce, jobs and the degree of centralization of activities. Perhaps less obvious in its effects on activities is legislation regulating competition, coordination and planning of transport services. Since the change of government to Conservative in Britain in 1979 there has been a change in emphasis towards competition which has made planning and coordination more difficult. The Transport Act 1980 abolished the need for licensing by the Traffic Commissioners of coach services on routes over 50 kilometres in length. Licences for other services have generally been granted unless it could be shown that they would be against the public interest. The onus of providing evidence was therefore reversed. No longer could the Traffic Commissioners regulate fares except to protect the public against abuse by monopolies. The Transport Act 1985 abolished the need for licensing of local bus services of 15 miles or less. It also allowed for the reform of local transport undertakings as companies separate from the local authority, although local authorities may continue to own them. In this respect, Britain is moving towards the arrangements in several West European countries including West Germany and France.

Under the *Personenbeförderungsgesetz* (Passenger Transport Act) the West German Federal Government has delegated administration of public transport services to the *Länder* in consultation with the *Gemeinden* and transport operators. It is becoming more common for licensing to be dealt with by *Verkehrsverbünde* (Transport Associations) in which the Federal Government, *Länder* and municipal transport operators are represented.

In France, the *Loi d'orientation des Transports Intérieurs* (Interior Transport Guidelines Act) *1982* set out significant changes reflecting the decentralization measures favoured by the socialist government elected in 1981. It requires access to public transport at reasonable price for all users – the 'right to transport for all', as it has often been interpreted. All passenger transport operators must meet government specifications. The State and *communes* are responsible for ensuring a public transport service in urban areas. The *département* carries out this function in rural areas.

The *Loi d'orientation des Transport Intérieurs* also allows an authority responsible for public transport, after consultation with relevant *communes,* to prepare a *plan de déplacements urbains* (plan for urban journeys). Transport authorities are not compelled to prepare such plans, but in practice they are necessary in order to obtain subsidies. As well as public transport, these plans include policies for the private car, bicycle and pedestrian. Unlike public transport plans in Britain, *plans de déplacements urbains* must go to public inquiry. As their names imply, the French plan is concerned with a whole range of factors which affect journeys in urban areas including social factors and urban development. They have also been concerned to regulate all traffic flow and promote public transport by giving it priority in traffic.

Like the *plan de déplacements urbains*, the *Generalverkehrsplan* (Traffic Master Plan) in West Germany is a significant medium for considering together private and public transport networks, including cycleways and pedestrian routes. They are not statutory plans and have varied in form. The *Generalverkehrsplan* for Frankfurt (Frankfurt am Main Dezernat Planung 1976 and 1984) for example, is concerned with the evaluation of alternative transport networks in terms of a series of 52 objectives relating to accessibility, the environment, traffic restraint, city form, operation and cost. It is much closer in form to a land use/transportation study or structure plan than a plan for the management and operation of public transport.

Conclusions

Seen as three packages, the full range of planning and transport legislation is not very different between the three countries. The purposes of plan forms differ between countries but seen together they are concerned with similar issues. Some of the most significant differences have arisen only recently since the coming into force of the Transport Act 1985 and the abolition of the metropolitan counties and the Greater London Council. Britain now has no metropolitan local government.

Of the three countries, Britain certainly has the strongest planning profession, if not the only one. In France and West Germany (and most other countries), planning is still largely in the hands of architects, civil engineers, surveyors, lawyers and other professions. Planning as a separate profession is weak, especially in France. Perhaps because the old-established professions are firmly entrenched in France, in some respects planning has more force there. Britain has no equivalent legal definition of land rights as is found in the *Bebauungsplan* and *plan d'occupation des sols,* but is perhaps none the poorer for that.

In comparing public transport in Britain, West Germany and France, the difference in levels of subsidy is immediately striking. As the following chapters indicate, farebox receipts are commonly around 80% of operating costs in Britain, whilst 40% to 50% is common in France, and even less than 40% is not unusual in West Germany. Although there are differences in accounting procedures (considerable parts of French and West German 'subsidy' is really the result of taxes), British cities are nevertheless lowly subsidized. Since the Transport Act 1985 the differences seem likely to grow. The availability of taxes in West Germany and France whose proceeds are earmarked for public transport has coincided with much greater capital investment in *métros*, *U-Bahnen* and other urban railways than in Britain which will have promoted centralization and will have helped to prop up the economy of the city centres.

The weakening of control over local public transport routes with the Transport Act 1985 in Britain seems likely to have an effect on town centres. Some fares on well-used routes, mostly into town centres, may well come down in price, promoting centralization and strengthening city centres.

4 Physical planning principles

Structures of city centres

The problem is how to give easy access whilst protecting the environment, the chief threat being that of traffic. Generally, easy access and a peaceful environment are in conflict, except where there are underground railways, or perhaps even surface railways and trams. In smaller cities of less than about 50,000 population, access by foot will be significant and in many West German (and other European) cities, access by bicycle is taken as a serious means of preserving the environment. Many French and British cities however, were altered in the 1960s and early 1970s to meet the demands of the private car. In many cases, little or no consideration was given to the bicycle and now, whether there is demand for it or not, extensive alterations would be necessary for safe cycling to be possible in city centres.

As soon as massive urban road construction started, not all planners were convinced of its unmitigated blessings (see the plan for Redditch in Chapter 10 for example). In Britain, however, it was not until the early 1970s that the concern for the protection of the environment began to take precedence over urban road building, and in France some urban motorways are still under construction. By this time, most British cities had acquired sections of a ring road (and a few cities including Birmingham and Coventry, a complete one) fed by high capacity radial roads. Public transport outside London and Paris consisted of buses with severely declining numbers of passengers with a few local rail lines converted from the national network.

By the mid-1970s the reduction or near-abandonment of urban road construction in favour of environmental management was accompanied by an awareness that public transport had more of a part to play than just being a social obligation to those who did not have a car. Marseille, Lyon, Tyne and Wear, Lille and Nantes have all opened *métros* since 1977 and several other British cities such as Sheffield, Manchester and Birmingham are seriously considering one.

There can be no doubt that public transport contributes to the prosperity of city centres by bringing in workers, shoppers, customers for leisure pursuits and a whole range of other activities. It seems equally certain that the city centre would be much the poorer if planning policies became so tough on the private car that shops, leisure facilities and perhaps even offices were dissuaded from being there. It is as well not to forget that one of the attractions of many suburban shopping centres is their easy access by car and parking.

Principles

Design of a public transport network depends on representing the interests of several groups – passengers, operators, people and businesses in the locality and taxpayers who provide subsidy. This translates into three main groups of concerns:

1 accessibility of passengers; spread of routes to give a maximum walking distance of 300 m, 500 m or whatever policy is decided upon, avoiding as far as possible long detours and retracking which result in time wasting and expense of operation;
2 minimization of nuisance in terms of noise, fumes, visual intrusion to people in the locality;
3 operation of routes within the technology available – slopes, turning spaces for buses, traffic conditions. Cross-city routes have been unpopular with operators due to heavy traffic in the town centre.

Several case studies examining the application of such principles, mostly in the suburbs of French cities, are contained in *Les transports collectifs dans l'aménagement des quartiers nouveaux* (Centre d'Etudes des Transports Urbains 1978a). These principles will also apply in city centres, but with some differences in emphasis.

Maintaining access whilst protecting the environment has been a fundamental concern in city-centre transport planning. Several principles have been used to try to obtain the benefits of both. Generally this has been based on defining two kinds of area: one in which access takes priority over the environment, and another in which the environment takes precedence.

1 Ring road or primary roads peripheral to the city centre with parking (Fig. 4.1). There are limits to what is a reasonable walking distance from the car parking near the ring road and this means that this arrangement works best in cities with a tightly

defined inner ring road, such as Birmingham. In larger cities, public transport could be provided from the car parks, but the city centre would have to be all the more attractive to overcome this drawback.

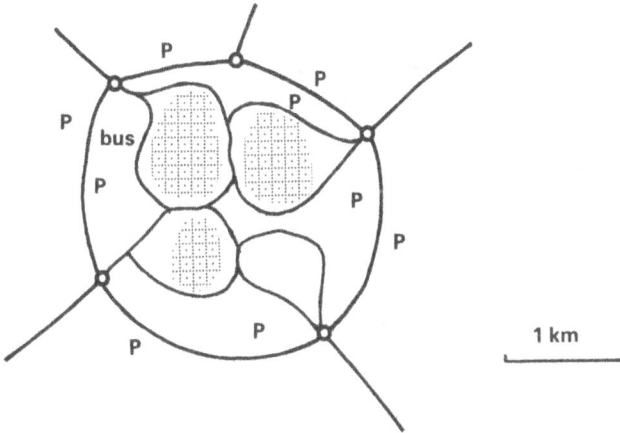

Fig. 4.1 Suburb-city centre bus orientated structure. Environmental areas may be shopping streets.

2 Public transport goes right into the centre to keep it alive (Fig. 4.1 and 4.2). Services between suburb and city centre may terminate in a loop. Cross-city services, usual for trams, light rapid transit and *métros* but not so common with buses, give a much better service to

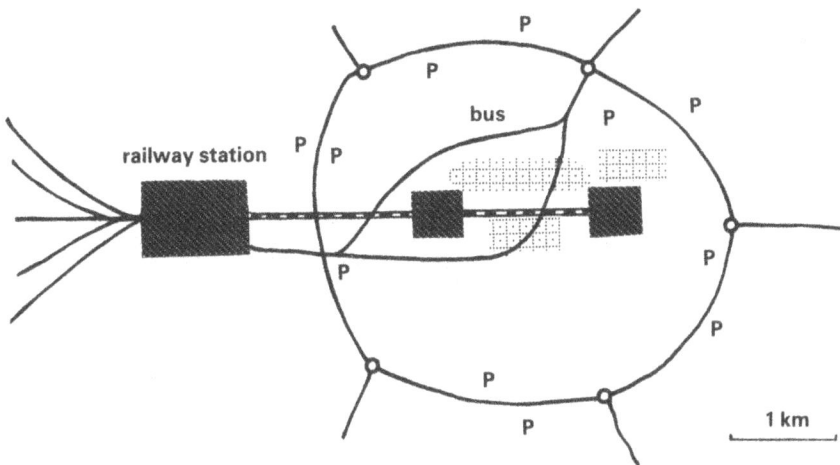

Fig. 4.2 Public transport node peripheral to the city centre with a *métro* spur projecting into environmentally protected areas.

the opposite side of the city centre. A problem is uneven demand from suburbs at opposite ends of the route. Services terminating at the opposite side of the city centre are a solution where the road pattern allows. Although these services should provide a better service to passengers, they have not been popular with transport operators due to heavy traffic in the centre.

3 The city centre may be divided into cells with access from an inner ring road. To prevent through traffic there may be no connection between the cells. Birmingham, Frankfurt, Hamburg, Lübeck, Bremen and Redditch all adopt this principle.

4 Collecting nodes for public transport peripheral to the centre – often developed around a railway station. These are common in France and West Germany, for example in Lille, Bremen and Lyon (Part Dieu). Often these are on or just outside and inner ring road or partial ring, for example in Hamburg, Lübeck and Frankfurt. These nodes collect through traffic to avoid it having to cross the city centre. Local public transport services connect node with city centre (Fig. 4.2).

5 Cross-city *métro*, tram or light rapid transit associated with environmental improvements such as pedestrianization or conservation of historic buildings (Figs 4.4 and 4.5). Busways could serve a similar function and in fact offer some advantages over rail transport except where demand is very heavy. Busways are less expensive to construct, more easily converted to general-purpose roads if required, more easily adapted to changes in passenger demand and fit more easily into a suburban bus network (Niblett 1972).

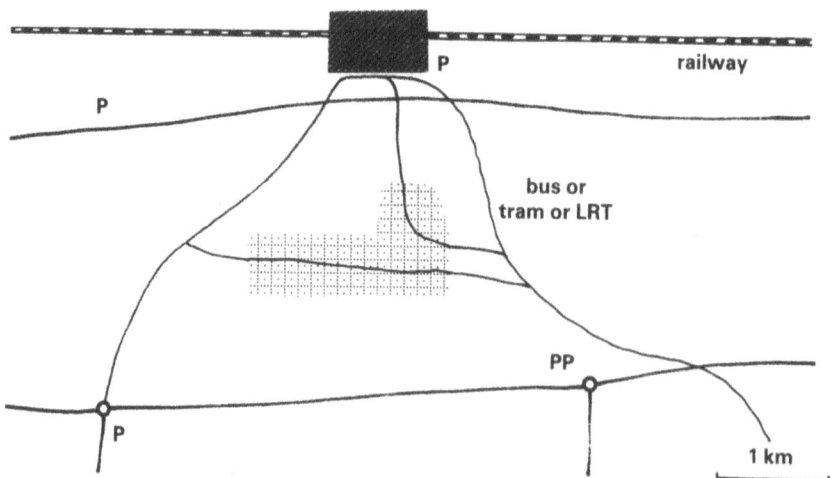

Fig. 4.3 Public transport node peripheral to the city centre; penetration into environmentally-protected areas is by train, metro or light rapid transit.

6 Buses converge on suburban *métro* stations and terminate there. To permit change between buses it is usually better to terminate at the *métro* terminus and a few other stations rather than evenly at all stations. Some buses are therefore kept out of the city centre resulting in a significant reduction in the level of traffic on a few streets.

7 Satellite city centre. To filter off some potential journeys to the city centre and escape high accommodation costs, routine office jobs

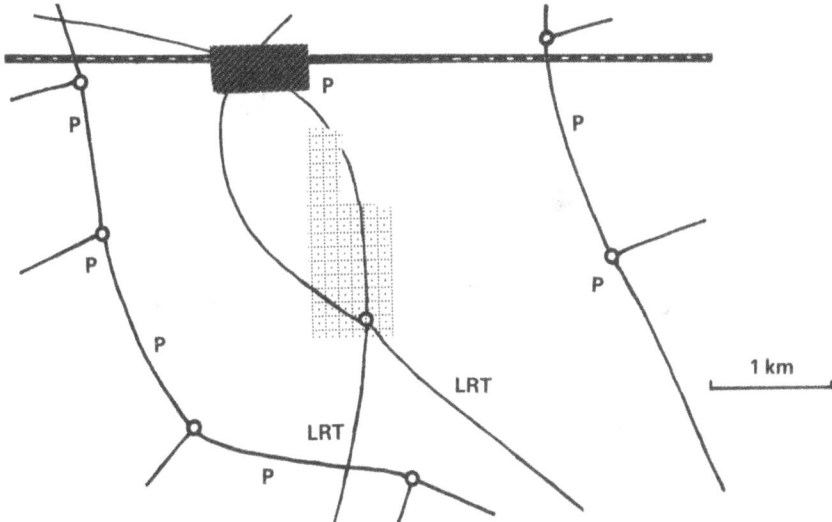

Fig. 4.4 Two *métro* lines form a circle in the city centre; main line railway station at one of them (cf. Marseille, Lille).

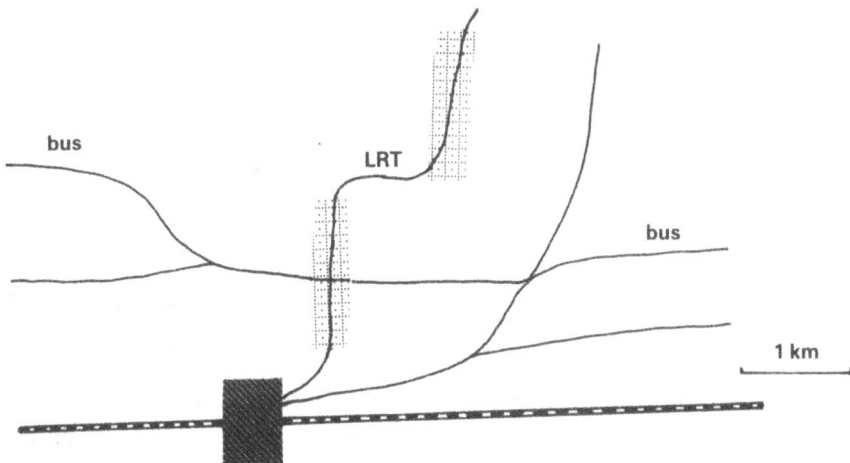

Fig. 4.5 A *métro* replaces a busy bus route with limited pedestrianization and other environmental improvements above.

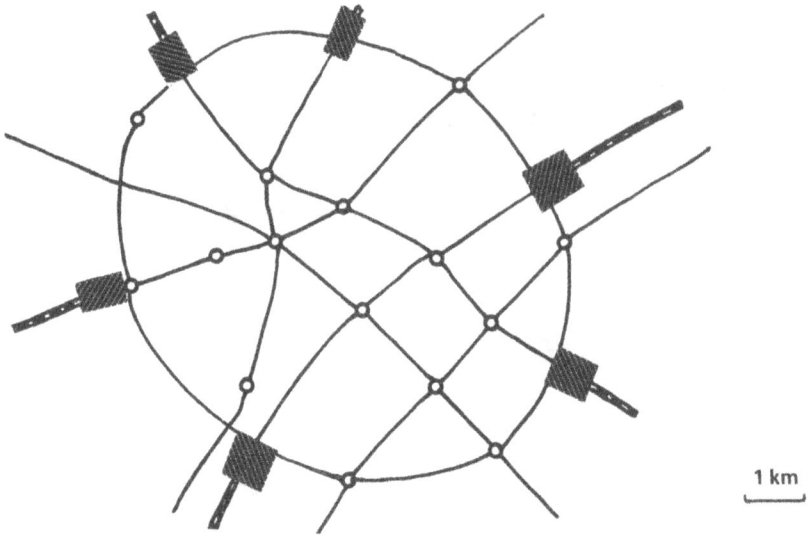

Fig. 4.6 A large city centre with several main line railway stations on a *métro* circle. Other cross-city *métro* lines mostly pass through at least one main-line station.

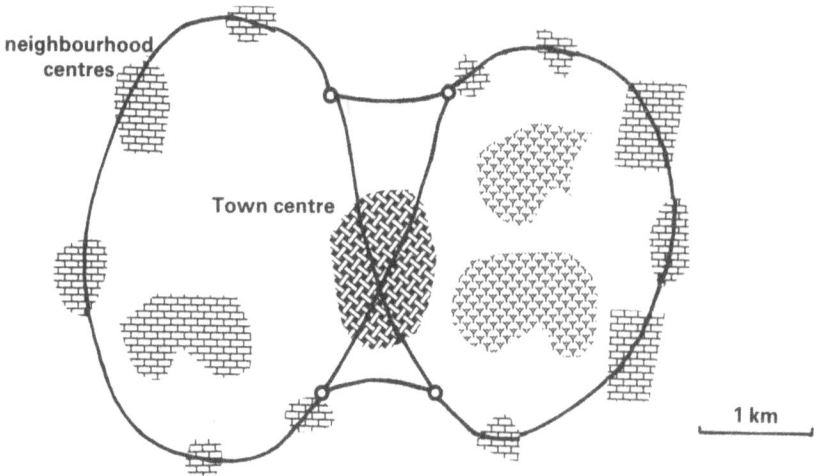

Fig. 4.7 The figure-of-eight viable only for smaller cities. The rather circuitous route to the city centre from the suburbs would render distances too great in larger cities.

which do not require immediate contact with the city centre have been relocated to satellite city centres. Some of these office developments are within a few kilometres of the city centre e.g. West End (Frankfurt), La Défense (Paris), Elephant and Castle

(London), Hamburgerstrasse (Hamburg), whereas others are physically and functionally more separate, e.g. Croydon (London), Niederrad (Frankfurt). City Nord (Hamburg), although only six kilometres north of the city centre on one of the eight city growth axes, more closely resembles a peripheral office development in its spacious design.

Land uses and density

Accessibility is a very powerful factor in determining land prices, profitability and intensity of use. This is one area where planning authorities have good reason not to try to divert the property market but to harness it to raise local taxes. So many city-centre railway stations such as New Street in Birmingham and Perrache in Lyon are now buried amongst revenue-producing shopping centres and office blocks or close to very large office developments such as the Frankfurt *Hauptbahnhof* near the banking centre, Victoria or London Bridge stations in London or the Gare de Lyon in Paris.

Workers seem to be more conscious of time spent travelling than are shoppers. Hence offices tend to take precedence around city-centre railway stations, especially in the largest cities such as London. The effect is much less clear in Paris, where long-distance commuting has never occurred on the same scale as in London. Offices also offer scope for much greater density of building and can take greater advantage of accessibility.

Shoppers are more readily put off by obstacles such as lifts, staircases or the need to cross busy roads than are those travelling to work. Hence the popularity of shopping on ground floors, first floors and basements. Much of the effort towards environmental improvements and particularly pedestrianization in city centres is in shopping streets. Pedestrianization has nearly always been accompanied by an increase in turnover in shops which demonstrated the value of the environment to shoppers and the economic significance of it to traders (and municipalities). Rail-based public transport is not essential for pedestrianization but it certainly helps to reduce the unpleasant effects of congestion, stopping and alighting from buses on the periphery of pedestrian areas and can give large-capacity direct access with little environmental change – especially if it is underground.

Typically there will be some shopping very close to the transport node, perhaps in underpasses leading to it or on the ground floors of high-rise office blocks. Further away, density decreases and the environment takes precedence; pedestrianization is common perhaps in a conservation area.

Bus stations

Bus stations allow passengers to change easily between services. Well designed in the open air they can also be an attractive gathering place for shops, leisure and other facilities. Badly designed, they can be a dirty, smelly, oil-soaked environmental disaster as usually seems to be the case when they are squeezed under the corner of a shopping centre, multi-storey car park or office block. Properly planned, they occupy a lot of valuable land, too valuable for a bus station in the eyes of many large city authorities. Perhaps that explains why many small towns such as Burnley (Chapter 11) or Redditch (Chapter 10) have got a much better bus station than most large cities, including Birmingham for example (Chapter 5).

The alternative to bus stations is to use only kerbside stops. A wide street such as Corporation Street in Birmingham or a bus-only street such as High Street (Birmingham again) can be used to act as an unofficial linear bus station. Many large cities rely on these, apparently with no great urgency for a bus station. Where journeys are mostly between suburbs and city centre there will be no great need for change between services and therefore no pressing need for a bus station. There may be a greater need for suburban bus stations, especially where the opening of a *métro* has created collecting points for bus routes at the stations, and which at the same time can act as exchange points between bus services.

Priority for public transport

To increase access to the city centre and/or reduce environmental damage, most large towns and cities in Western Europe have, since the 1960s, made some attempt to give priority to public over private transport. Providing a new public transport network on its own track, such as a *métro*, is an expensive solution, generally not very effective in enticing travellers away from their cars. It may also be environmentally destructive if not underground. It is much cheaper and easier to give priority to public transport on the existing network as a supplement or substitute for a new network.

There are many variations on bus lanes and busways using existing roads with varying degrees of separation from the general traffic circulation. Designs are reviewed in *Les aménagements des axes prioritaires de transports collectifs* (Centre d'Etudes des Transports Urbains 1979a). As a minimum, such priority may be only a few tens of yards of bus lane at a congested road junction to help reduce delays to public transport. At the other extreme it can be a separate road network for exclusive use by buses. There are examples of each at Birmingham and Redditch.

Priority for buses at traffic lights has also been used. A bus may trigger off a switch perhaps 150 m or 200 m before the lights to cause them to change to green, although this has sometimes caused intolerable problems for the rest of the traffic.

As well as increasing the speed and reliability of public transport, it is important to give information to passengers waiting at stops. Many transport authorities do not even take the elementary step of having timetables at all or even most stops. At the other extreme, some have installed electronic devices which show the location of buses along a route and therefore give an indication of how long there will be to wait to the next bus – very useful for doing a few quick errands nearby rather than wasting time at the bus stop.

Where trams share the road carriageway, they encounter some of the problems of buses, perhaps to a lesser extent because of their greater weight, strength and accumulated momentum. Nevertheless, many tramway networks have been put underground through the most congested districts for example in Marseille and Lille (Fig. 6.18). In Bielefeld, although nearly two-thirds of the 25 km of tramway routes are on road carriageways, since 1970 significant sections have been developed on their own reserved track and in the city centre, underground. There are plans to separate more of the tramways from the general circulation thus converting step by step what was a *Strassenbahn* into a *Stadtbahn* or light *métro*. The current situation and plans are reviewed in Centre d'Etudes des Transports Urbains 1978b. In Karlsruhe, by 1978, more than two-thirds of the tramways were already on reserved tracks, but the City had recently abandoned a plan to put the central part of the network underground due to high cost (300 to 500 million DM at 1978 prices). Instead, further improvements of surface track and further priority at traffic lights were envisaged (Centre d'Etudes des Transport Urbains 1978b).

Pedestrianization

The majority of pedestrian schemes are associated with predominantly shopping uses. This presumably reflects the importance of the environment for browsing around shops at greater length than the usually direct journeys to a place of employment and the ease with which shoppers can move their custom to another locality if dissatisfied. However, there are pedestrian precincts associated with other uses – civic and cultural, such as around many cathedrals, and educational, as in many universities.

The elimination or strict regulation of traffic is nearly always accompanied by an increase in trade on the streets concerned. Even nearby streets, which may be expected to suffer by a neighbour becoming more attractive, seem to benefit in the majority of cases (Roberts 1981, Chapter 7). Nevertheless, except for the unlikely event of pedestrianization creating increased spending, the transfer of custom can only be from further afield.

Pedestrianization is more likely to transfer custom than create it, but transfers can be quite valuable to a local authority if the positions of their boundaries are favourable. A more persuasive argument for pedestrianization is that if it is sufficiently attractive to cause customers to transfer from elsewhere, then it seems likely that both new and old customers are deriving greater benefits from their shopping expeditions than before. As economists would say, both are enjoying an increase in consumer surplus. Although most of these benefits will not appear in any financial accounts, many municipalities and others may feel them to be a quite legitimate support for pedestrianization.

It seems clear that the restriction of traffic results in greater benefits to those using or working in the streets affected. In economic and environmental terms the effects are comparable in kind if not in extent, to the building of a *métro*. How the benefits compare in size is open to question. The costs of pedestrianization are certainly less. It may be that pedestrianization is not only complementary to improvements in public transport but is also an alternative means of progressing to the same ends.

Pedestrian access to public transport

Distance to walk is not the only significant consideration. 400 yards along an interesting shopping street or a pleasant river bank seems far less than the same distance along a boring underpass or cheek-by-jowl with heavy, fast-moving traffic. Not enough thought has been given to the distance pedestrians must walk and even less thought has been given to the quality of their surroundings as they do so. All too often, urban road-building has turned what was once a short direct walk into a series of irritating zig-zags and confusing changes of level of twice the distance or more, even if you manage to avoid taking the wrong route. As well as the frustration and boredom for the pedestrian, it does not make sense to waste their shoe leather walking up and down ramps and underpasses if they could be passing interesting and revenue-producing shopping frontages. In many West German and other European cities, pedestrianization has been accompanied by careful design of pavements, street furniture and lighting to enhance the environment.

Shops and underpasses can be symbiotic. Underpasses are potentially well-trodden paths; shops can take advantage of their passing trade and indeed enhance it. By acting as a source of surveillance they might help to reduce the chances of violent crime. Shops may even instill a sense of identity into these anonymous and, for the visitor, confusing paths.

5 Birmingham and Coventry

Birmingham

The civic literature is in the habit of describing Birmingham as Britain's second city. Although many commercial organizations have chosen Manchester as their second centre after London, indeed the administrative area of Birmingham does contain more people than any outside London (or even London itself since the disappearance of the Greater London Council on 1st April 1986). Within the 264.3 sq. km, Birmingham had a population of 1,006,527 in 1981, a decrease of 8.33% since 1971. It is by far the largest of the seven districts within the West Midlands County, which within the 899.43 sq.km had a population of 2,648,939 in 1981, a decline of 5.17% since 1971. Administrative areas do not often correspond closely to social and economic realities, trade and other catchment areas. Birmingham is closer than many other large cities in this respect. Parts of Sutton Coldfield, within Birmingham administratively, probably have affinities with Birmingham no stronger than Solihull, Dudley or Walsall and weaker than parts of Sandwell, all of which are outside the municipal area of Birmingham. Birmingham is thus the largest centre of a conurbation of 2,335,000 population (all the West Midlands except Coventry), although Wolverhampton and to a lesser extent Walsall, Dudley, Sutton Coldfield and Solihull all have a lot of functional independence.

Compared with some West European cities, Birmingham is a city of uniformity. It has neither the poverty nor the conspicuous wealth to be seen in many continental European cities. Only four of the 42 wards are occupied at a density of more than 6,000 persons per square kilometre and none as high as 10,000. Only four are occupied at less than 3,000 persons per sq. km. It is a city of municipal standards which, on the one hand, have saved it from squalor, but on the other have resulted in a lot of standardized housing surrounded by standardized, least cost landscaping. Even the site is unremarkable. In the 1960s, much of the city centre was redeveloped with priority being given to traffic circulation over the built environment. The city centre is remarkably accessible and traffic congestion relatively low. Due to quite successful separation of busy roads and footpaths, pedestrians do not have to suffer the overbearing intrusion of heavy traffic but they do have to put up with rather circuitous routing and confusing changes of level. Parts of Birmingham city centre are not at all friendly to the visitor on foot. Redevelopment has left it with relatively few landmarks. The few landmarks that remain are easily

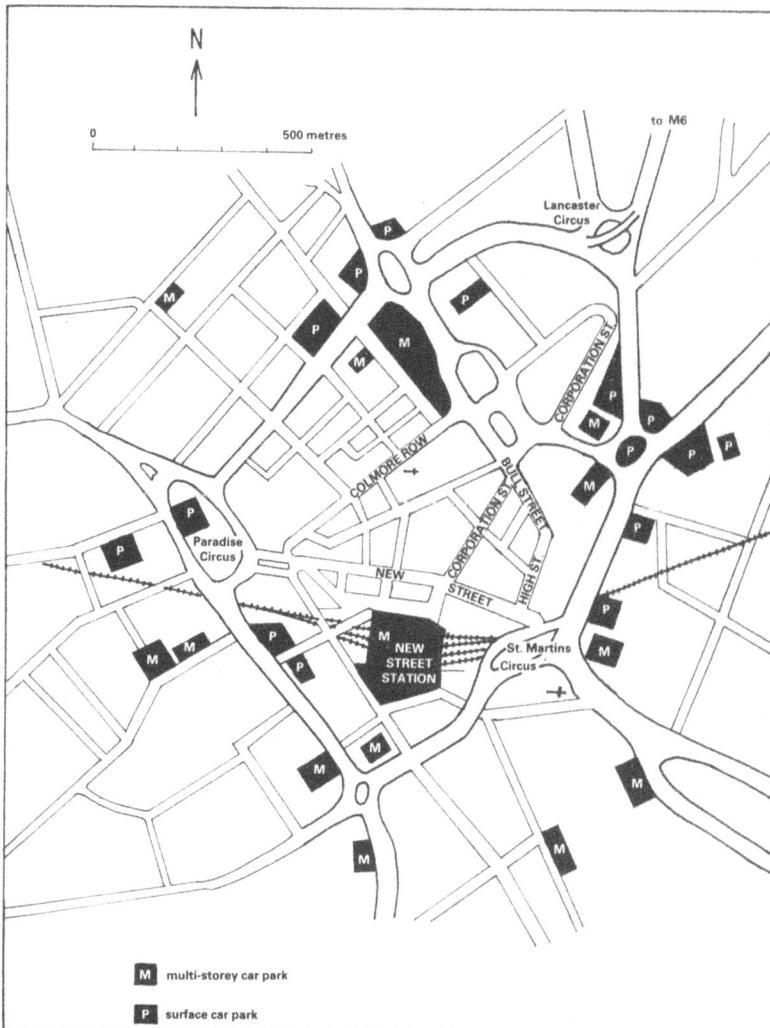

Fig. 5.1 Birmingham city centre.

missed when the pedestrian is channelled down an underpass, forced to go in several unwanted directions and then, not having a compass, emerges totally confused as to direction.

The city centre has been accused of being shabby, lacking in high quality shops and litter-strewn, of being poor environmentally and of lacking leisure facilities (Beaufort Research 1984). It is true that there are no significant outdoor recreation facilities. There is no park, only a few small areas of open space and no river or lake which could form the centrepiece of a leisure area. Shopping suffers from competition from other large West Midland towns such as Wolverhampton, Walsall and Dudley, and large suburban centres such as Erdington and Northfield.

Fig. 5.2 Birmingham city centre is remarkably accessible. The Aston Expressway links the M6 in the distance to the inner ring road.

Birmingham Central Area local plan (Birmingham City Council 1980, 1982 and 1984) reflects these criticisms in adopting as one of its main aims the upgrading of the environment. Other principal aims are the consolidation of the shopping centre, expansion of the office quarter and restructuring of the inner industrial ring. A fundamental concern of the plan is to create confidence in Birmingham as a commercial centre and to reverse decline by improvement of the infrastructure. Better access is needed to the main retail area of Corporation Street/High Street/New Street and so too is better pedestrian access across the inner ring road. More car parking is seen

Fig. 5.3 The inappropriately named Paradise Circus, which is almost a bus station, where the inner ring road cuts through the civic centre.

Fig. 5.4 Colmore Row, one of the main stopping points for buses in the city centre.

as part of the needed improvements in infrastructure. Plans for six new multi-storey car parks were approved by the City Council in late 1985. Being publicly financed, these are a reversal of the policies of the structure plan (West Midlands County Council 1980), taking cognizance of their effects in supporting the prosperity of the city centre.

The Central Area Plan covered a wide area. In 1981 this had a population of 22,000, two-thirds of which was in the inner suburbs of Ladywood, Lee Bank, Highgate and Newtown. The city centre had only a small population. There are only a few blocks in the centre of the city – council high-rise blocks at Holloway Circus and Stephenson Tower above New Street Station and the Aston University residences.

In 1976, 157,000 were employed in the city centre, a decline of 6% since 1971 (Table 5.1).

Table 5.1 Employment levels in Birmingham city centre.

	%	% change 1971–76
Services	30	+ 1.5
Manufacturing	27	−21.0
Public administration	18	− 6.0
Retailing	9	−11.0
Transport and distribution	8	
Others	8	

Fig. 5.5 There are 15,280 car parking spaces in Birmingham city centre, of which 9,966 are within or immediately adjacent to the inner ring (right). Of these, 82% are multi-storey. Aston University is in the background.

Thus employment in the city centre had been relatively stable whereas there had been a large decline in the manufacturing districts just outside the centre.

Perhaps the most important proposals in the infrastructure improvements in the Central Area local plan have been the proposals for a light rapid transit network, developed by the West Midlands County Council in the years immediately before its abolition (Tym and Partners 1983, West Midlands County Council/West Midlands Passenger Transport Executive 1984). Passenger flows along the routes were expected to be in the range of 5,000 to 8,000 passengers per hour, which would only be one-third or less of the demand needed to justify the heavier forms of underground such as in London or Paris. Altogether, ten lines have been proposed radiating out from Birmingham, with a total length of 54 km and at a cost of £500 million. An initial phase of four lines is proposed at a cost of £210 million. Lines from the northern suburb of Kingstanding and West Bromwich to the west of the city centre approach Snow Hill Station and then tunnel underground to New Street Station. Lines from Sutton Coldfield to the north and Chelmsley Wood to the east meet at Gosta Green on the north-eastern fringe of the City Centre and then cross the city centre, under Broad Street and out to Five Ways 1 km to the west. All the tracks in the city centre were to be underground, except a small section approaching Snow Hill Station. The reasons for the proposals are largely economic and environmental. In particular

Fig. 5.6 The Town Hall (left) and Council House, New Street.

the light rapid transit will connect with important development sites at Snow Hill where 51,000 sq. m of offices have recently been completed, Paradise Circus (180-bed hotel, 21,330 sq. m of offices and cultural facilities under construction) and Broad Street where a major convention centre for the whole city is proposed. The proposals are also seen as an important part of the city's effort to improve the city centre environment and in particular the pedestrianization of New Street and Corporation Street. At the time of writing ten months after the abolition of West Midlands County Council, the likelihood of the network being built seems uncertain.

Structure of the city centre and public transport

The inner ring road encloses an area of only 80 hectares. Some city centre uses spill over in the west around Paradise Circus (Repertory Theatre, Baskerville House and the Registry Office, for example) and in the south around the Bull Ring (Bull Ring Centre, markets, Digbeth Coach Station, for example) but the centre is still small for a city of Birmingham's size.

The city centre is cut into two unequal parts by Priory Queensway/Snow Hill Queensway. North-east of this are the hos-

Fig. 5.7 Bull Street.

pitals and the court/legal district. The larger part to the south-west comprises three main districts: the civic area around Paradise Circus, the business area on Colmore Row and to the shopping area centred on New Street, Corporation Street and High Street.

There are no traffic measures which enforce a cellular pattern of circulation within the centre although many of the bus routes follow such a pattern. For example, many of the buses from the south-west enter the City Centre from Broad Street, follow a circuit along New Street, Corporation Street and Colmore Row and leave again via Broad Street. Comparable circuits can be traced from other directions using High Street, Bull Street, Priory Queensway and Snow Hill Queensway, for example.

Pedestrianization is mainly confined to some of the minor streets within the area bounded by New Street, High Street, Bull Street and Colmore Row and in the civic area. The Birmingham Shopping Centre above New Street Station and the Bull Ring Centre are also pedestrian, and Bull Street and High Street are bus-only. The circuses on the inner ring road are pleasant if you can forget that you are encircled by a traffic roundabout. Apart from Old Circus, they are not used as much as they might be. Despite being encircled by heavy traffic, Lancaster Circus and St Chad's Circus particularly are quite isolated within and are not on main pedestrian routes. They attract a lot of tramps which although fulfilling a possibly useful purpose does tend to deter other city centre users. So, of the main shopping streets, only New Street and Corporation Street are open to traffic (and there are plans to close parts of these [Birmingham City Council 1980, 1982 and 1984]) and several of the side streets leading from Corporation Street are pedestrian-only.

According to an Aston University survey (see Appendix), two-thirds of the journeys to the city centre are by public transport (66% weekday on-peak, 65% off-peak and 70% on Saturdays). Public transport in the city centre is largely by bus. According to the survey, 77% of weekday peak hour public transport journeys are by bus (23% by train), 84% weekday off-peak journeys are by bus (16% by train) and 80% on Saturdays (20% train). Two local railway lines from the south-east end at Moor Street Station, the other six and the long-distance lines go to New Street.

Of the 118 West Midlands Travel bus services which mainly serve Birmingham, 81 are from suburb to centre, 30 from suburb to suburb, four cross-city and three are circular. Ninety-five services enter the city centre. There is no local bus station. The main stopping points are on Corporation Street (29 services), Colmore Row (28), New Street (23) and Bull Street (15). There must be a great deal of pedestrian movement between these points and the many other stops scattered throughout the centre. Twenty-one Midland Red services also enter

Fig. 5.8 The inner ring road at the junction with Broad Street. Some city-centre uses spill over outside the inner ring (left) on the way to Five Ways.

at the Midland Red Bus Station. These serve towns in the Midlands outside Birmingham.

Most of the buses do not go to the opposite side of the city centre from which they entered. There are therefore quite considerable distances to walk for those destined for the opposite side of the city centre from where they came.

A public opinion survey on the potential for public transport in the city centre

In the mid 1980s Birmingham city centre faces two main issues for which public transport may seem to offer some alleviation or partial solution. Firstly there is the decline in prosperity of shops, office uses and other commercial uses. So far in the 1980s two of the four department stores in the city centre have closed. One (the Co-op) is being re-developed and the former Debenham's premises were reopened late in 1985 as Hamley's toy shop. Lewis's and Rackham's remain on Corporation Street. Some parts of the shopping centre, particularly on the north-eastern end of Corporation Street, have proved difficult to let and other parts such as in the Bull Ring Centre have not attracted prosperous, high-rent premises as was hoped. Some offices have remained vacant for considerable periods. In short, there is scope for strengthening of the city centre in economic terms. Improved public transport, as may result from a new light rapid transit network, has been seen as a significant way of strengthening the economy of the city centre, by providing an attractive alternative to the use of the car to visit suburban centres for shopping, leisure and cultural activities, and as a way of at least maintaining the attraction of the city centre for employment.

A second main issue in the city centre where public transport may contribute is the environment. With an area of only 80 hectares within the inner ring Road, Birmingham city centre is small for a city of its size. Development is dense and activities intense. Pedestrian congestion and lack of open space and pedestrianized streets are problems which may contribute to the unattractiveness of the centre in the eyes of many people, who at present choose to shop in relatively wealthy suburbs such as Sutton Coldfield and Solihull. A light rapid transit network would certainly be a help to further pedestrianization. It might also result in some reduction in general congestion, if those attracted to use it are not replaced by others travelling by car into the city centre. If they are, then more people will be brought in for a given level of congestion and this may help satisfy economic if not environmental objectives.

A survey on the potential for public transport in the city centre was carried out at Aston University in November 1985. 476 city centre users were selected randomly and interviewed at 34 locations distributed throughout the area within the inner ring road to represent business, shopping, legal and other districts.

As expected, those travelling in on optional journeys on Saturdays and for shopping expressed a greater willingness to increase their journeys if light rapid transit was built. So far, only present city-centre users have been contacted. Perhaps less expected was the lack of difference in attitudes between present public and private transport users. There seems to be some potential for capturing journeys from the private car.

Coventry

Coventry has a long history as an important industrial town and this is still reflected in the buildings and structure of the city centre. By the eleventh century it had become the fourth-largest city in England, based on the weaving of wool and, later, silk. Many of the surviving historic buildings date from the fourteenth and fifteenth centuries when Coventry was at the height of its medieval prosperity.

The weaving industry remained until the 1860s when the protection of import taxes was removed by the government of the day, at a time which also coincided with competition from Lancashire. Soon afterwards, one of Coventry's other main industries, watchmaking, also declined due to foreign competition. The city lay in decline until the 1880s when the basis of twentieth century manufacturing industry was laid, at first in the form of sewing machine and bicycle production. Firms such as Hillman, Riley and Singer were originally associated with bicycles and attracted related industries such as Dunlop in 1890.

The motor industry expanded rapidly after its beginnings in 1896. Between then and the 1960s population rose rapidly from 70,000 in 1901 to 336,746 at the time of the 1971 Census. Employment in Coventry was particularly attractive during the depression years of the 1920s and 1930s and the two world wars acted as stimuli to its industries. By British standards bombing during the second world war was severe. After destruction in November 1940 two symbols of

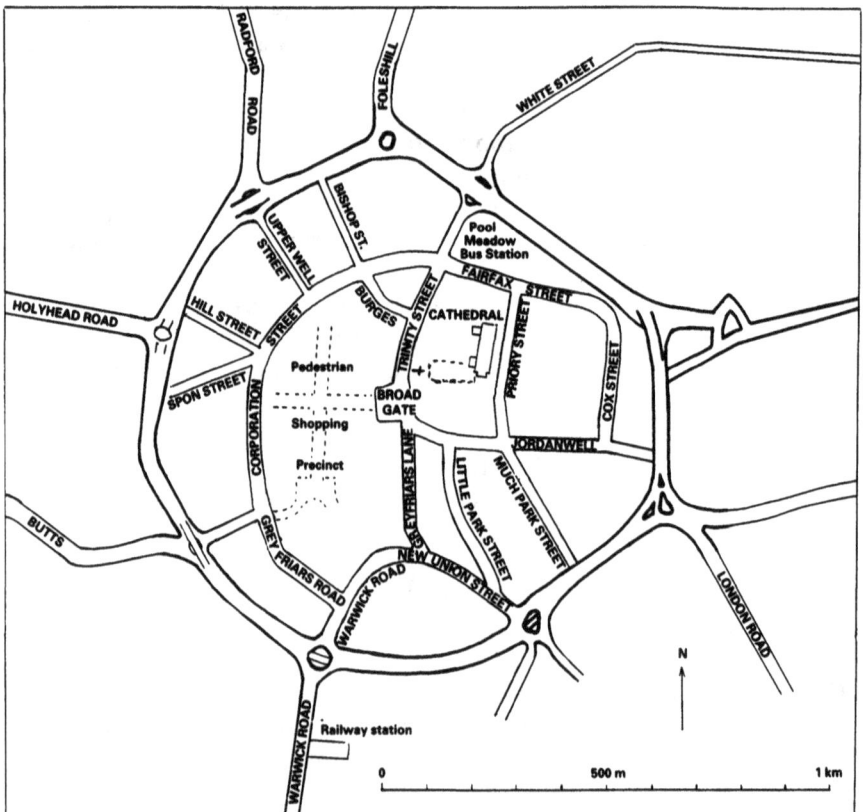

Fig. 5.9 Coventry city centre.

Fig. 5.10 Pedestrian precinct of the 1950s; the spire of the old cathedral on the opposite side of Broadgate can be seen in the background.

Fig. 5.11 Inside the remains of the old cathedral.

Fig. 5.12 Pedestrian area between the old and new cathedrals.

reconstruction in particular became associated with the City: the pedestrianized shopping area and, on the opposite side of Broadgate, the new cathedral and the pedestrian precinct around it. Both remain major elements in the structure of the city centre.

Throughout the twentieth century up until the early 1970s, Coventry was characterized as being a prosperous city with a high proportion of skilled manual jobs in the motor vehicle and aircraft industries, and low unemployment. In 1975 British Leyland, Chrysler, GEC and Rolls-Royce together employed 62,800 people in the city. These industries have been severely affected by the economic recession and rising unemployment since the mid-1970s.

At the eastern end of the former West Midlands Metropolitan County, Coventry is physically separate from the remainder and is relatively independent, although some of its main industries are characteristic of the region. At the time of the 1981 Census the 9,654 hectares within the area of Coventry City Council, which closely corresponds to the built-up area, had a population of 313,815. This was a decrease of 7.3% since 1971, after an increase of 5.8% between 1961 and 1971 when the motor car industry was more prosperous. 26.9% of heads of households were in semi-skilled and unskilled occupations (socio-economic groups 10 and 11) compared with 18.5%

Fig. 5.13 Pedestrian shopping precinct built shortly after the Second World War. Development generally and shopping in particular are not so dense as in Birmingham. The 90 hectares within the Coventry Inner Ring Road house 85,930 sq. m of shopping floor space, whereas the 80 hectares within the Birmingham inner ring contain 197,320 sq. m.

in Great Britain, and 15.2% were employers, managers and in the professions (SEGs 1 to 4) compared with 14.9% in Great Britain. 42.7% of households had no car compared with 49.5% in Birmingham and 39.5% in Great Britain as a whole.

Table 5.2 Mode of travel to work in Coventry and Birmingham.

	% Car	Bus	Train	Foot	Other
Coventry	47.7	24.1	1.0	16.5	10.7
Birmingham	44.1	33.9	3.0	12.5	6.5
Great Britain	50.2	16.0	5.7	15.7	12.4

Source: 1981 Census of Population.

Public transport in the city centres

Forty-six West Midlands Travel bus routes serve the Coventry district, and all except one, which serves Meriden and Balsall Common in the green belt to the west, enter the city centre.

Table 5.3 Bus services in Coventry and Birmingham.

	Suburb to centre	Suburb to suburb	Cross city	Circular
Coventry	31	1	15	0
Birmingham	81	30	4	3

The greater size of Birmingham has no doubt influenced the development of purely suburban routes not entering the city centre and at least the outer circle if not the inner circle route in Birmingham. Coventry public transport is even more centralized than that of Birmingham. All the Coventry cross-city routes enter within the inner ring road. The lack of cross-city routes in Birmingham may reflect the greater density of development in the centre which is a disincentive to bus travel there and probably also reflects the greater size of the city as a whole. Longer cross-city routes make timekeeping more difficult, especially when travelling through the centre.

Table 5.4 Modal split of journeys into the centres of Birmingham and Coventry, morning peak, 1983.

	Bus	**Rail**	**Private (%)**
Birmingham	37.9	11.2	50.9
Coventry	33.1	1.6	65.2

Source: West Midlands Passenger Transport Executive, Annual Statistical Report 1983–1984, Market Research and Management Services Unit, Birmingham, 1984.

Whereas in Birmingham the modal split for peak hour journeys into the city centre is roughly 50:50 public to private, in Coventry it is approximately two thirds private. This reflects the greater market for public transport in Birmingham and the generally better services there (Table 5.5).

Table 5.5 Bus services in the South and East Divisions of WMPTE 1983/84.

	Bus km (millions)	Population (1981)	Bus km/ person
South Division (Birmingham)	61.2	1,006,527	60.8
East Division (Coventry)	11.2	313,815	35.7

Source: West Midlands Passenger Transport Executive, Annual Statistical Report 1983–1984, Market Research and Management Services Unit, Birmingham, 1984.

Fig. 5.14 Broadgate is one of the main bus termini in the city centre.

Unlike Birmingham, Coventry has a central bus station, Pool Meadow, with 30 of the 46 services. All of the other services except two stop at the Burges or on Broadgate between the main pedestrian shopping area and the pedestrian precinct centred on the old and new cathedrals. Many of the Pool Meadow services also stop on Corporation Street which functions like part of a ring within the Inner Ring road. So all of the routes serving the Coventry district except that connecting the outlying villages of Meriden and Balsall Common stop within a very short distance of each other in the city centre. Whereas many of the routes serving Birmingham city centre terminate in the form of a circle enclosing perhaps one third or one quarter of the area within the inner ring on the same side of the centre as the direction from which the service came, the services in Coventry do not circulate to the same extent. Broadgate and Burges are very close to the geographical centre of the area within the inner ring and distances to walk to other parts of the centre are not great. Pool Meadow Bus Station is somewhat off-centre, but many of the services to it stop elsewhere in the centre.

As well as the 46 West Midlands Travel services, there are also 21 Midland Red services to neighbouring towns in the Midlands, and all these services go into the city centre.

Fig. 5.15 Pool Meadow bus station.

Conclusions

Both Birmingham and Coventry have gone much further towards providing accessibility by road to the city centre than almost any other large British cities. Both have used the principle of providing good access to the inner ring road where this consideration has taken precedence over the environment. Within the inner ring some priority has been given to the environment, far more so in Coventry than in Birmingham. Birmingham centre gives the unmistakable impression of having been planned by road engineers, whereas in Coventry

Fig. 5.16 Many of the Pool Meadow bus services also stop on Corporation Street.

urban design and the environment have clearly had more priority. Coventry centre relies heavily on the buses to support activities and land uses, and Birmingham, probably due to its greater size, does so even more. The high density of building development in Birmingham centre creates problems for accommodating buses. Sufficient space has not been available for a bus station and some of the streets, particularly High Street and Colmore Row function as such with loss of road space to other users and considerable environmental intrusion, especially in High Street.

Roads were given high priority in the redevelopment of the city centre from the 1960s. Birmingham is one of the few British cities where it is possible to drive right to the inner ring road without leaving roads of motorway standard. It is well supplied with nearly 10,000 off-street public parking spaces within or adjacent to the inner ring and 95 bus services penetrating inside the inner ring. With only 32 railway stations in the whole of Birmingham there can be few districts where the train is more convenient than the bus for a journey to the city centre. The fact that more than 20% of journeys to the city centre are by train shows some preference for rail over bus travel. The survey of city centre users supported this.

Also given high priority in the development of the city centre has been rateable value. The Birmingham Shopping Centre above New Street Station has been remarkably successful despite the limited local railway network. The connected but less accessible Bull Ring Centre is noticeably less successful.

Environmental considerations have been fitted in between, underneath and around the roads. Temple Row and connecting passages and, on the opposite side of Corporation Street, Union Street and its environs are areas where the pedestrian has had priority. The circuses are a worthy but not entirely successful attempt to make use of traffic roundabouts as open space. Old Square contains useful shopping, but the rest are not used as much as they might be.

Within the constraints set by these priorities and with very little investment in public transport, Birmingham city centre has been successful. Some parts are quite run-down, notably Broad Street, Corporation Street between Old Square and Lancaster Circus and, to a lesser extent, the Bull Ring Centre, but plausible attempts are planned to alleviate the worst problems.

6 Lyon, Marseille and Lille

These are the three largest cities in France after Paris. In recent years, all three agglomerations have had a fairly stable population total but decline in the inner areas. The city centres as defined below in Cahiers Français (1985) are much more widely defined than elsewhere in this book. Actually these figures refer to the main *communes* of each of the four conurbations. The city of Marseille covers a much larger part of the conurbation than do Lyon or Lille, hence the lower rate of decline.

Table 6.1 Population figures for main French cities.

	Unités urbaines			City centre	
	No. of communes	1982 pop.	% change 1975–82	1982 pop.	% change 1975–82
Paris	335	8,707,000	+0.4	2,176,200	−5.4
Lyon	84	1,220,800	0.0	413,100	−9.6
Marseille	28	1,110,500	+1.3	874,400	−3.8
Lille, Roubaix and Tourcoing	55	936,300	0.0	366,900	−8.7

Unlike Marseille, Greater Lyon and Lille are administratively *communautés urbaines*. That for Lyon comprises 55 *communes* of which Lyon (413,095 population in 1982) and Villeurbanne (115,960) are by far the largest in terms of population. The *communauté urbaine* of Lille comprises 54 *communes* of which Lille, Roubaix and Tourcoing are the largest. The City of Marseille on the other hand is a single *commune* of 874,000 population.

Both Lyon and Marseille are on quite spectacular sites of some physical difficulty for building. The centre of Lyon is now on a

Fig. 6.1 Lyon city centre looking eastwards from the basilica on Fourvière: the cathedral and medieval old town in the foreground, River Saône and peninsula in the middle distance. In 1982 55,000 people lived on the Peninsula compared with 95,000 in 1962.

Fig. 6.2 The northern entrance to Perrache, the main station for long distance and local public transport in Lyon.

peninsular extending northwards from the confluence of the rivers Rhône and Saône, with a recent extension to the centre about one kilometre to the east (Part Dieu). The medieval old town is to the west at the foot of the very steep hill of Fourvière which is the site of the Roman nucleus. Marseille, by far the largest port in France, is constrained between the sea and the hill masses towards north, east and south and has been notorious for its congestion, partly as a result of site restrictions. Practically the whole of the 23,000 hectares or so that are suitable for building have been developed.

All three cities have had large scale dispersal of housing during the past 20–30 years with consequent problems for public transport (Syndicat des Transports en Commun de la Région Lyonnaise/Agence d'Urbanisme de la Communauté Urbaine de Lyon 1983). *Grands ensembles* were developed at high density from about 1960 onwards, relying on buses for public transport (Simpson 1987). Jobs in the city centre and inner city have not declined in the same way as population. Thus in Marseille there has been an increase in jobs both in the centre, from 205,000 in 1962 to 235,000 in 1975, and on the periphery, from 79,000 in 1954 to 119,000 in 1975 (Mazzella 1981). In fact Marseille is still very centralized. One-third of the jobs are located on only 1.3% of the area of the city (Tuppen 1980). The *métro* has been seen as a way of spreading out employment e.g. to Castellane on the fringe of the city centre, and reducing the journey to work to the city centre.

Public transport networks

All three cities have conurbation-wide transport authorities with uniform transferable ticketing systems and with separate *métro* companies. Those for Lyon and Marseille are explained in Chapter 2.

Table 6.2 Comparison of *métros* in Marseille, Lyon and Lille.

	Opened	Line	Length (km)	Stations	% of public transport journeys
Marseille *métro*	1977	1	9	12	} 19 } 28
	1984	2	6	9	
Lyon *métro*	1978	A	9.4	13	
		B	3	6	} 31
		C	3	5	
Lille *métro*	1983	1	13.5	18	30.6

The public transport networks in each city have a lot in common. Each relies on a limited *métro* in the centre and some inner suburbs, and, in Lille, extending out eastwards to the new town of Villeneuve d'Ascq. The *métros* are fed by buses from the middle and outer suburbs. Considerable numbers of buses from districts not served still penetrate into the city centres.

The *métros* have been seen as a way of reversing the declining fortunes of the city centres and inner areas in particular, as a way of joining the suburbs to the main areas of employment and of providing interchange facilities with the railways. In Marseille there have been large numbers of immigrants from North Africa (Kinsey 1979) and particularly severe problems in the city centre.

The Lyon *métro* is routed through a more densely built-up area than the Marseille *métro*. There is a population of 123,000 living within 500 m of Line A (Société Lyonnais des Transports en Commun 1983, p.44) which implies a population density of 12,870 persons per sq. km. From information in Ville de Marseille (1983a), it is possible to calculate the population densities around the Marseille *métro* stations. The areas around the Line 1 stations average 8,101 persons per sq. km, and around Line 2 stations 6,736 persons per sq. km.

In Marseille the *métro* has been used as a series of collecting points for bus routes, especially in the south and north-east of the city, and very few buses from these districts go into the city centre. Most that do go into the city centre are from the north and east of the city. As Bouffartigue (1985) points out, the Marseille *métro* serves areas dependent on central area employment instead of the working-class districts of the north and east and suburb-suburb journeys. Thus it has helped to prop up the city centre and accentuate the difference between employment in the centre and residence in the suburbs.

The *métro* has been particularly valuable in reducing the need for buses in Marseille city centre, which has few wide roads suitable for large vehicles and where traffic speeds are slow. However, it is doubtful whether it has reduced city-centre traffic congestion. One transport official summed up the motorist's view as follows: 'Well done. You have built a *métro* to reduce the number of motorists in the city centre. Now I can use my car to go into town'. Mazella (1984) came to similar conclusions in relation to the Marseille suburbs of St Antoine and St Loup: public transport has had little effect on the use of private cars for journeys to work in Marseille as anyone who has one available will use it. However, this effect of *métros* is not without benefit. In Lyon, it has been observed that the greatest increase in trade in the city centre following the opening of the *métro* came from the north-west of the city which is not served by it. The *métro* vacated space for these people to travel in by car (Watel 1985).

Nevertheless, the *métro* has coincided with a revival in the use of public transport. The annual number of journeys at around 130 million in the mid-1980s is close to what it was in the mid-1950s, having declined to 80–90 million between 1968 and 1977. As well as the *métro*, tariff reform, marketing strategy and reorganization of bus routes along with the metro have no doubt contributed to this revival. Despite this, subsidies have continued to increase from nothing in 1965 to 15 million francs in 1968 and 170 million francs in 1979.

The same phenomenon of using the *métro* to collect some bus routes in the suburbs and some revival of city-centre shopping and commerce can be observed in Lyon and Lille.

Fig. 6.3 The exit from Perrache leads into the main pedestrian area in the peninsula with the *métro* underneath.

In Lyon, the opening of the *métro* coincided with a reversal in the fortune of public transport and also a reversal in the downward trend in shopping turnover in the peninsula (Chambre de Commerce et d'Industrie de Lyon 1983; Watel 1985). There has been some stimulation of development outside the city centre too. In Marseille too, the *métro* can be seen to influence development. Between 1975 and 1978, 25% of new dwellings were built adjacent to Line 1. A science park is being developed adjacent to the north-eastern terminus of Line 1. Also, much housing has been converted to offices along the first *métro* line. Although the revitalization has not been as great as might have been at first expected, the *métros* in both Lyon

Fig. 6.4 High-density building is characteristic of Central Lyon as in many other continental cities. Despite a 23% decline in population since 1968, the *commune* of Lyon has a density of 9,030 persons per sq. km (1982).

and Marseille have been important in improving the images of the two cities and in promoting a closer relationship between city centre and suburbs. It is now not so much a case of directing and planning growth but of 'selling' the town (Dalmais and Mazzella 1985).

In Lille, Malabry (1985) felt that the *métro* had not been a decisive influence on the centre of Villeneuve d'Ascq which has been expanding with the new town but may have been more significant for Lille city centre. It has however, had some bad influence on the inner suburbs. It has, accelerated the closure of shops run by the elderly and has taken away young people for other activities. This fear was also expressed by Dalmais van Straaten (1985) in relation to the Lyon *métro*. Once shoppers are on the *métro* they might as well go right through a suburban centre such as Gratte-Ciel to the city centre. Arrival of a *métro* station is unlikely to reverse urban decline but might accentuate urban growth which is already taking place. On the other hand, where a *métro* station is also a change of means of transport, it may have a beneficial effect on shopping around the station. Working in Lyon suburbs, Sanson (1984 and 1985) found changes of transport means had a considerable effect on local shopping, especially in persuading shoppers to call back to the centres at times other than the time of transport exchange. The *métro* seems to have become more competitive to the private car for shopping than for other journey purposes. In a survey reported by Malabry, 25% of *métro* users had a private car available at the time of their *métro* journey.

All three *métros* are presently being extended. Line 2 of the Marseille *métro* is due to be extended by a further 2½ kilometres with three stations in 1987, and further extensions are envisaged beyond the other present-day termini. In Lyon, Line C is being extended northwards, and a new line D of eight kilometres and linking Lines A and B is expected to be in service in 1988. A second line in Lille from the main railway station, north-westwards to Lomme is under construction. With Line 1, it will form a ring around the city centre, rather like the Marseille *métro*, although the ring is rather larger. The two connections with Line 1 will be at the main railway station and Porte des Postes (also a junction on the *Périphérique*).

Fig. 6.5 La Duchère. *Grands ensembles* were developed in Lyon from the early 1960s, relying on the buses for public transport.

Table 6.3 Percentage of public transport journeys.

	Lyon	Marseille	Lille
Bus	68	68	58.6
Métro	31	28	30.6
Trams	—	4	10.8

The streets in Marseille tend to be narrow and winding, in fact, characteristically Mediterranean. On 800 out of 1000 km of streets, the carriageway width is less than 7 m and this has resulted in very low traffic speeds. It has also accentuated the need for priority measures for public transport. The average speed of the buses is only 13 km per hour and much less than this in the city centre. Since the 1975 *Plan de Transport* four main groups of policies have been followed (Lissarague 1984):

1 Half of the investment in public transport has been on forms using

Fig. 6.6 Part Dieu, Lyon, has been developed primarily as an administrative and cultural centre (*cité administrative*, Communauté Urbaine de la Région Lyonnaise [COURLY]), concert hall, library and commercial centre. The Credit Lyonnais tower is on the right. It reflects decentralization from Paris (Tournier, Laferrère and Thomas 1973) as well as from Lyon city centre. In 1984 the shopping turnover of Part Dieu was 1,400 million francs compared with 2,500 million francs in the city centre (Watel 1985).

Fig. 6.7 The centre of Marseille, viewed from the south-east.

Fig. 6.8 Marseille Vieux Port. Congestion here hinders north-south traffic through the city centre. Three-quarters of the traffic entering Marseille is from the north.

their own track – *métro*, tramway, *SNCF*.

2 Best use has been made of existing infrastructure, peripheral parking, tariff reform, busways and priority at traffic lights.

3 Road improvements – ensuring continuity of capacity, feeders to the *métro*, have been implemented.

4 Expressways – A7 *(Autoroute Nord)* ringway, connection to *Autoroute Nord* and *Autoroute Est* in tunnel – are priorities from the proposals in the *SDAU*.

In Marseille there is also one remaining tramway extending from the *métro* station at Noailles three kilometres eastwards to St Pierre with nine stops and accounting for 4% of public transport journeys. In Lille, the Mongy tramway connects the three cities of Lille, Roubaix and Tourcoing.

Structure of the city centres

In none of the three cities is there an inner ring road. In Lille there is a complete *périphérique* 3 km in diameter enclosing the city centre and most of the western half of the *commune*. There are no urban motorways within it, but there are several wide, straight *boulevards* unfortunately converging on *ronds points* (cf. Paris, Chapter 9). In Lyon, the *périphérique* is broken in the west by the rivers and the hill of Fourvière. The Lyon *périphérique* is 6 km in diameter with the city centre at the western fringe. Within the street pattern is closer to grid-iron than *boulevard* and *rond point*. Marseille has nothing resembling a complete *périphérique*. There is an incomplete motorway box about 4 km in diameter around the inner city. The *Autoroute Nord* penetrates right into the city centre to within 0.5 km of the Vieux Port, and the expressway, Avenue du Prado, terminates on the edge of the city centre at Castellane. Each ends at a *rond point*, connected to each other right across the busy streets of the city centre and bisecting La Canebière, one of Marseille's main shopping streets.

In all three cities therefore, the road network does little to reduce traffic penetration into the city centre, except to filter off some through traffic at the *périphérique* (and even this function is far from complete in Marseille). Partly due to topography in Lyon and Marseille, and due to the absence of a suitable route without great disruption in all three, there is no road designed specifically to act as a distributor within the city centre. The overall planning of the structure of the city centre has been much less ordered and more piecemeal than has been the case in all German cities in this book (and Birmingham too).

Fig. 6.9 New shopping centre Bourse in the heart of Marseille.

Fig. 6.10 The Marseille *métro*. Three kilometres of Line 1 is on the surface or elevated, but only two-thirds of a kilometre of Line 2.

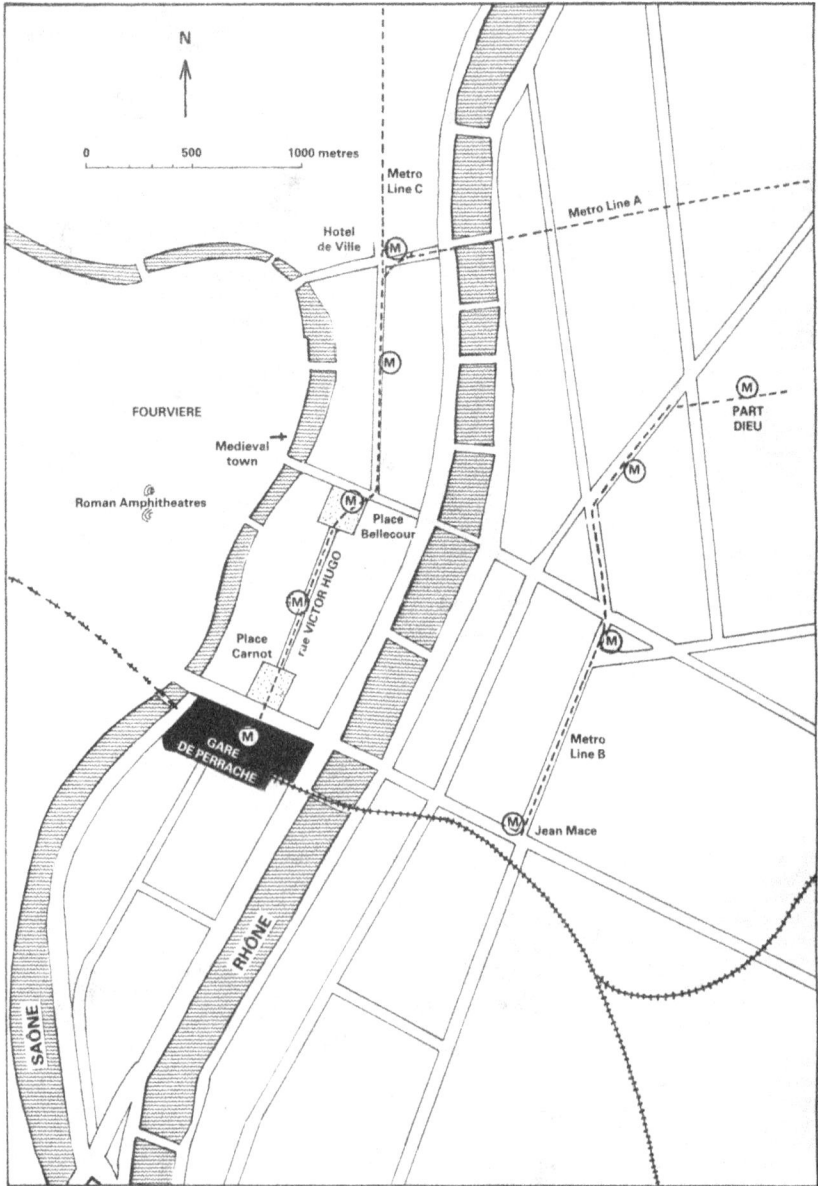

Fig. 6.11 Lyon city centre.

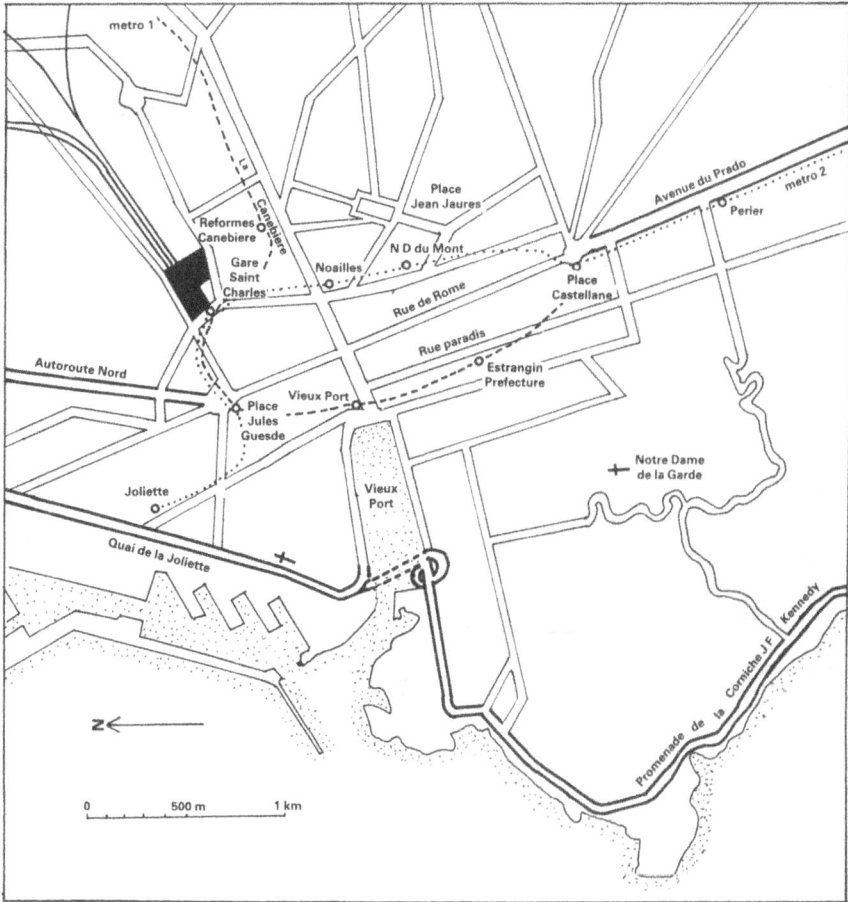

Fig. 6.12 Marseille city centre.

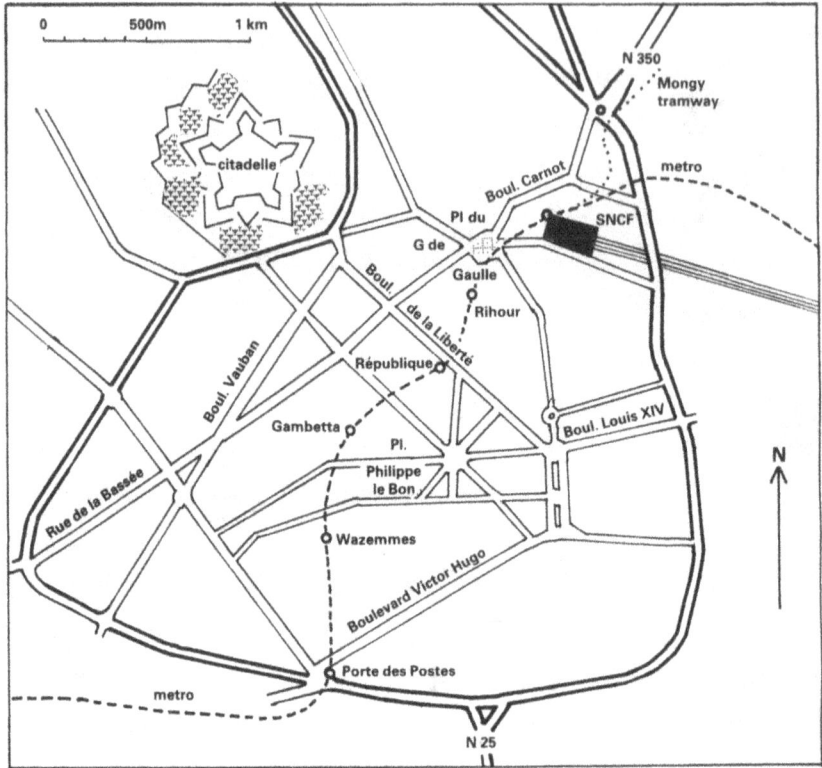

Fig. 6.13 Lille city centre.

Fig. 6.14 The main railway station in Lille is also the centre of the local public transport network for buses, Mongy trams and the *métro* (both underground here).

Fig. 6.15 High density housing in the centre of Lille.

Fig. 6.16 République *métro* station, Lille.

Fig. 6.17 Pedestrianization above the *métro* in Lille.

Fig. 6.18 The Mongy trams connect the three main towns of Lille, Roubaix and Tourcoing. Here the tram has just emerged from the tunnel on the fringe of Lille city centre on the way to Roubaix.

Conclusions

The extent of adaptation of the city centre to changing needs, principally that of the private car, has depended on three main considerations:

1 The availablility of physical space and constraints on building and environmental management. These French cities have not had the advantage of a ring of former fortifications around the present city centre to adapt as a ring road as has happened in many other European cities.
2 The balance between the environment and road building. Urban road-building usually means sacrificing the environment in certain areas to enhance it in others. The French solution has been to avoid this destruction more than has happened in many West European cities including all the German cities in this book (and Birmingham). The Lille *périphérique* is the best example of an attempt to filter off city centre traffic, but its distance from the centre limits its effectiveness in this respect.
3 The choice between individual property rights and collective well-being. French planning has generally placed a lot of emphasis on individual rights, which would have made unacceptable the disruption and large-scale demolition necessary for inner urban roads.

With finance from *versement transport* (Chapter 3) the *métros* have been developed to save the city centres from economic decline, to revitalize inner city districts and to improve the city centre environment. The *métros* have proved to be modest competition for the car. They may not have reduced the amount of road traffic in the city centre, as the small numbers who have abandoned their cars to travel on the *métro* may have been replaced by other motorists.

However, the *métros* have probably allowed more people to gain access to the city centre and have given some increase in freedom of movement. In Lille, one third of *métro* journeys were not previously undertaken (12% in Lyon, 6% in Marseille) and in Marseille and Lyon, one third of the *métro* journeys were previously not by public transport. Although not essential for the limited pedestrianisation which has been achieved, they have helped it. The benefits of *métros* may not have been as great as was hoped, but neither have the costs been as clear as has been published. The true costs of a capital project, when we consider taxes, subsidies and opportunity costs, for example, are very ambiguous, particularly in times of high unemployment.

The local authorities have been willing to invest in capital works to address the problems that have faced them during the last two or three decades. The high levels of subsidy to public transport at 61.1% in Lyon and 51.8% in Marseille compared with 19.6% in the West Midlands (*Société des Transports en Commun de la Région Lyonnaise* 1984b, *Régie des Transports de Marseille* 1984a, West Midlands Passenger Transport Executive 1984a) might imply that *métros* are a very expensive way of providing public transport.

The *métros* are not the only important attempts to address local planning problems. Part Dieu in Lyon is developing into what is almost a second city centre. Like the British tower blocks of flats, the *grands ensembles* have had their problems, but perhaps should be compared with what they replaced. The *Ville de Marseille* has achieved a great deal during the past 30 years *(Ville de Marseille* 1983). Although climatically and environmentally it is a very favoured city, it still has grave problems associated with crime and immigration The *métro* quite suddenly becomes deserted in the evenings. The street begging and squalor give the impression of this City not belonging to the prosperous countries of Western Europe. The 280 km which separate Marseille from Lyon span a lot of social and cultural differences. So too do the even shorter distances which separate Marseille from the coastal resorts to the east. Marseille is not part of the French Mediterranean playground socially or economically.

7 Hamburg and Frankfurt

Hamburg

Hamburg, the largest second-tier city in West Germany, like Munich, Frankfurt, Cologne, Düsseldorf, Stuttgart and other regional metropolises, has taken on more than second-tier functions since the annexation of Berlin. Situated on the River Elbe, 70 km from the open sea, it has remained the largest port in West Germany, despite much of its pre-Second World War hinterland now being in the Deutsche Demokratische Republik. Like West Berlin and Bremen, Hamburg is a separate *Land* (State). The boundary of the *Land* is quite tightly drawn around the built-up area and a considerable part of its hinterland for journeys to work, shopping and other functions is in the neighbouring *Länder* of Schleswig-Holstein and Lower Saxony. This has made the steady loss of population in recent decades appear more than it actually is when we consider the functional area. The population of the *Land* has declined every year since 1966 from 1,850,600 to 1,609,500 at the end of 1983. The number of persons employed in the *Land* has declined from 980,000 in 1967 to 890,000 in 1984 (Hamburger Verkehrsverbund 1984). So whilst the number of residents has declined by 13% the number of persons employed has declined by 9%. Unemployment in 1984 was 11.2% of the workforce.

Hamburg has a large centre in relation to its population. The area within the inner ring road is three and a half times that of Birmingham and in Hamburg, some important city-centre uses spill over the inner ring. To the east is the *Hauptbahnhof* and Berliner Tor, an important office and administrative centre. To the south is the administration associated with the Port, and, to the west, perhaps Hamburg's second claim to fame after the Port, there are the entertainments of St Pauli and Reeperbahn. Hamburg is both a business and a leisure centre. The waterways from the Elbe to the Aussenalster attract visitors, and in between, business and leisure come together in the Fleets or canals flanked by offices and commerce.

Since the early 1920s the structure of development in Greater Hamburg has been based on a series of growth axes on road and rail transport (Albers 1977, Freie und Hansestadt Hamburg 1973). Since 1955 there has been a joint planning office with Schleswig-Holstein to secure the planning of these growth axes. Shortly afterwards, a similar arrangement was made with the *Land* of Lower Saxony. The overall planning of housing location and density and traffic structure has been much influenced by the Hamburg Density Model (Kruger, Rathmann and Utech 1972, Utech 1982). This combines three concepts: firstly axes of growth (eight of them extending into neighbouring *Länder* and some of them branching), secondly the structuring of the main transport network around rapid transit with bus feeders and park and ride and thirdly a hierarchy of central places along these axes. Maximum tolerable walking distances to

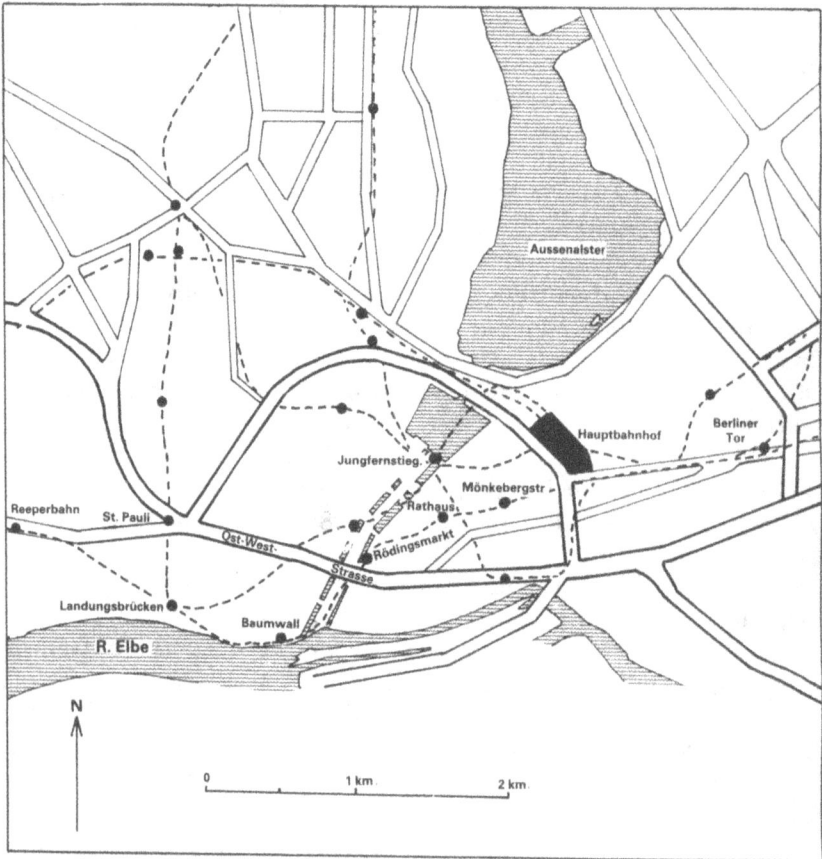

Fig. 7.1 Hamburg city centre.

Fig. 7.2 Hamburg Hauptbahnhof on the eastern fringe of the inner ring road is at the centre of the rapid transit network.

Fig. 7.3 City-centre uses extend to Berliner Tor, an important collecting point on the rapid transit network 1.5 km east of the inner ring road.

public transport and average housing density, each variable according to location, are specified in the model.

Three main concentric zones have been defined. In the outer zone,
the private car is to be dominant with buses to the rapid rail axes and
park and ride at stations. In fact, park and ride is provided at 48
rapid transit stations with a total capacity in 1981 of 7,250 places
(Hamburger Verkehrsverbund 1981). In the transitional zone there is
also park and ride and a more dense bus feeder network. In the inner
zone, which extends up to about 6 km from the centre, public
transport is mainly by rapid transit and the roads are designed
mainly for business traffic.

There are seven *S-Bahn* lines, three *U-Bahn* lines run by the
Deutsche Bundesbahn and the Hamburger Hochbahn Aktiengesell-
schaft (HAA) and three other railway lines run by the Eisenbahn-
Gesellschaft Altona-Kaltenkirchen-Neumunster (AKN), making a
total of 13 rapid transit lines with a total length of 287.7 km
(excluding common sections) and 181 stations. All the *U-Bahn* lines
and six of the *S-Bahnen* converge on the *Hauptbahnhof* on the north-
eastern fringe of the city centre. The other *S-Bahn* line runs
outward from Altona west of the city, and the three AKN railway
lines run outwards from the *S-Bahn* and *U-Bahn* termini (or near
terminus) to serve the further suburbs.

Buses are largely confined to the suburbs. This is reflected in the
high average speeds (22.3 km per hour for the whole Hamburger
Verkehrsverbund) and the wide spacing of stops (average 619
metres). There were 1,337 buses in operation in 1985, a large number
for a city with such an extensive railway network.

Public transport in the city centre is very largely the six *S-Bahnen*
and three *U-Bahnen*. The city centre really comprises 3.5 sq. km
within the old city walls, and is now contained within what amounts
to an inner ring road with some extensions of city centre uses on all
sides (Fig. 7.1). Six of the rapid transit lines converge on Berliner
Tor, a major office centre 1.5 km outside the inner ring to the east
(Fig. 7.3). All of these and the other three rapid transit lines in the
city centre continue to the *Hauptbahnhof* and five rapid transit lines
continue to the very centre of the city at Rathaus/Jungfernstieg.
Apart from Rathaus/Jungfernstieg there are only four other rapid
transit stations within the inner ring (Fig. 7.1), all quite minor,
although there are several others on the fringes. Ladungsbrücken is
an important sea passenger terminal in the south-west of the city
centre and acts as an interchange for rapid transit, although in a
more minor way than Berliner Tor in the east of the city centre. The
Hauptbahnhof is by far the busiest passenger terminal with 277,000
passengers per day in 1984, compared with Rathaus/Jungfernstieg
(106,000) and Berliner Tor (98,000).

Fig. 7.4 Jungfernstieg, the main rapid transit station inside the inner ring.

Fig. 7.5 Jungfernstieg *U-Bahn* and *S-Bahn* station adjacent to the Binnenalster lake.

Six kilometres north of the city centre and only 2.5 km from Hamburg-Fühlsbuttel Airport is City Nord, an office park developed since the mid-1960s on one of the development axes included in the Density Model. By 1980 it comprised 700,000 square metres of office space (compared with 1,860,000 in the city centre), and by 1985 it had 35,000 workers (Husain 1980, Baubehörde Hamburg 1985). This policy of relocating mainly routine office jobs which do not need a city-centre location to the fringe of the city-centre reduces city-centre transport demand, and there are comparable projects in Birmingham (Five Ways), Lyon (Part Dieu), Paris (La Défense) and many other cities. City Nord includes, of course, a busy *U-Bahn* station, which is in fact on the same line as the airport, Rathaus/Jungfernstieg and the *Hauptbahnhof*.

Despite the falling population, the total number of public transport journeys rose by 6.2% from 1967 to 593,300,000 in 1984 (Hamburger Verkehrsverbund 1984). The proportion by rapid transit has remained fairly steady at 56%. In 1984 the buses accounted for 43% and ferries the remaining 1%.

Public transport in Hamburg is a story of innovation (in organization, fares and timetables since the HVV became the first West German metropolitan transport authority in 1965), bold investment in infrastructure and a very sharp awareness of the interrelationships between public transport and land use planning. As well as the

Fig. 7.6 Mönckebergstrasse, one of Hamburg's main shopping streets, with department stores, the spire of the Petrikirche on the left and the Rathaus in the distance. The *U-Bahn* runs underneath this street.

Fig. 7.7 The *U-Bahn* is actually an elevated railway here at Rödingsmarkt and along the quayside (Fig. 7.8). The spire of the ruined Nikolaikirche is in the background.

Fig. 7.8 *U-Bahn* at Baumwall.

Fig. 7.9 Ost-West Strasse forms the southern section of the inner ring, here crossed by the *U-Bahn* (elevated) railway near Rödingsmarkt station. The spire in the background belongs to the Michaeliskirche.

Fig. 7.10 Some visual interest has been achieved in this multi-storey car park near the Hamburg *Hauptbahnhof*.

Fig. 7.11 High-density housing redevelopment just outside the inner ring to the south-west near Landungsbrücken.

Density Model, there is a very clear and effective strategy of peripheral collection of passengers in the city centre with sorting according to functional area, and only limited and necessary penetration into the very centre of the city. Despite what might appear to be a superabundance of public transport, 60% of operating costs are covered by fare revenue, or 65% taking into account compensation for reduced fares to the handicapped and children. Heavy capital investment has allowed public transport to compete successfully with the private car in the inner city, and has helped the city centre to maintain its activities in difficult circumstances due to the partition of Germany as well as the rise of the private car, decentralization of housing and other factors common in cities throughout the western world.

Frankfurt am Main

Frankfurt, the largest city in the State of Hesse, is situated on the River Main 30 km upstream from the confluence with the River Rhine. At the beginning of 1983 the City of Frankfurt had a population of 620,186 within an area of 247 sq. km. Like Hamburg and the majority of other Western cities, it has been in decline for several decades. In 1963 the population was 691,000 and the

Generalverkehrsplan (Traffic Master Plan) allows for 580,000 by the
end of the century (Frankfurt am Main Dezernat Planung 1976 and
1984). The *Generalverkehrsplan* also allows for Frankfurt to increase
further as what is already an exceptionally strong and influential
central place for a city of its size. Like Hamburg, it has gained capital
city functions as a result of the partition of Germany. It is West
Germany's financial capital and one of the leading European and
World centres of banking and commerce. It also has Europe's second
busiest airport after Heathrow. Frankfurt is a very strong employment centre. The population:jobs ratio is 1:1.4 (compared with 1:1.8
for Hamburg).

The overall spatial planning of the conurbation has been based on
principles comparable to the Hamburg Density Model, except that
development along rapid transit axes tends to be separated more by
open space forming satellite settlements rather than continuous axes.
The *Regionale Raumordnungsplan* sets out a very clear hierarchy of
central places located on the *S-Bahn* separated by open spaces away
from the City of Frankfurt. Both Hamburg and Frankfurt rely on *S-
Bahn* and *U-Bahn* as the framework of local public transport with
suburban passengers brought in by bus. In Frankfurt the *S-Bahn* is
particularly extensive (326 km compared with 148 km in Hamburg).
The Frankfurt *U-Bahn* extends over 40 km compared with 89 km in
Hamburg. In Frankfurt, the 349 buses have a limited role as feeders
to the light rapid transit, whereas in the Hamburg transport
authority there are 1,337 buses. There are also 249 trams in
Frankfurt (none in Hamburg.) (See Table 7.10.)

Table 7.1 Comparison of public transport in Hamburg and Frankfurt.

	Hamburg	Frankfurt
S-Bahn lines	9*	14
S-Bahn length (km)	148.3*	325.6
U-Bahn lines	3	5
U-Bahn length (km)	32.7	40.3
Other rapid transit lines	3**	–
Other rapid transit length (km)	50.0**	–
Buses	1,337	349
Trams	–	249

*Made up of 6 lines in common with *Deutsche Bundesbahn* (110.3 km) supplied by
direct current and 3 lines alternating current/diesel also in common with *Deutsche
Bundesbahn.*
**AKN 29.9 km, ANB 10.2 km and EBO 9.9 km.

Sources: Frankfurter Verkehrs- und Tarifverbund (1976 and 1984); Hamburger Verkehrsverbund 1985.

Fig. 7.12 The *Hauptbahnhof* is one one of the busiest north-south roads in Frankfurt which is also one of the main *Strassenbahn* routes. This is crossed at right angles by a pedestrian underpass, *U-Bahn* and *S-Bahn* (both underground) leading into the very centre of the City to the Hauptwache and Konstablerwache. A small bus station, the only one in central Frankfurt is in the *Hauptbahnhof* forecourt.

The structure of the city centre is comparable to that of Hamburg. It is mostly confined inside the former city walls which are now followed on each side by a one-way road to form what now functions as an inner ring road. There is some extension of city centre uses particularly eastwards through the banking quarter to the *Hauptbahnhof* which like the Hamburg *Hauptbahnhof*, is peripheral to the centre. All the *S-Bahn* lines (except number 9 which extends from Offenbach to Ober-Roden) lead there. Seven of the *S-Bahn* lines continue eastwards to the Hauptwache and Konstablerwache right in the heart of the City Centre where they connect with all the *U-Bahn* lines. Thus the *Hauptbahnhof* is the collecting point of the *S-Bahn* network, which serves the outer suburbs and surrounding towns up to about 50 km distant, and the Hauptwache and Konstablerwache are the focal points of the *U-Bahn* network, serving the City of Frankfurt and northern suburbs mostly less than 10 km distant. The *Hauptbahnhof,* Hauptwache and Konstablerwache are connected under the city centre which is thus relieved of much of what would otherwise be surface traffic. This main east-west axis is crossed by the main north-south axis of *U-Bahn* lines 1, 2 and 3, extending to the

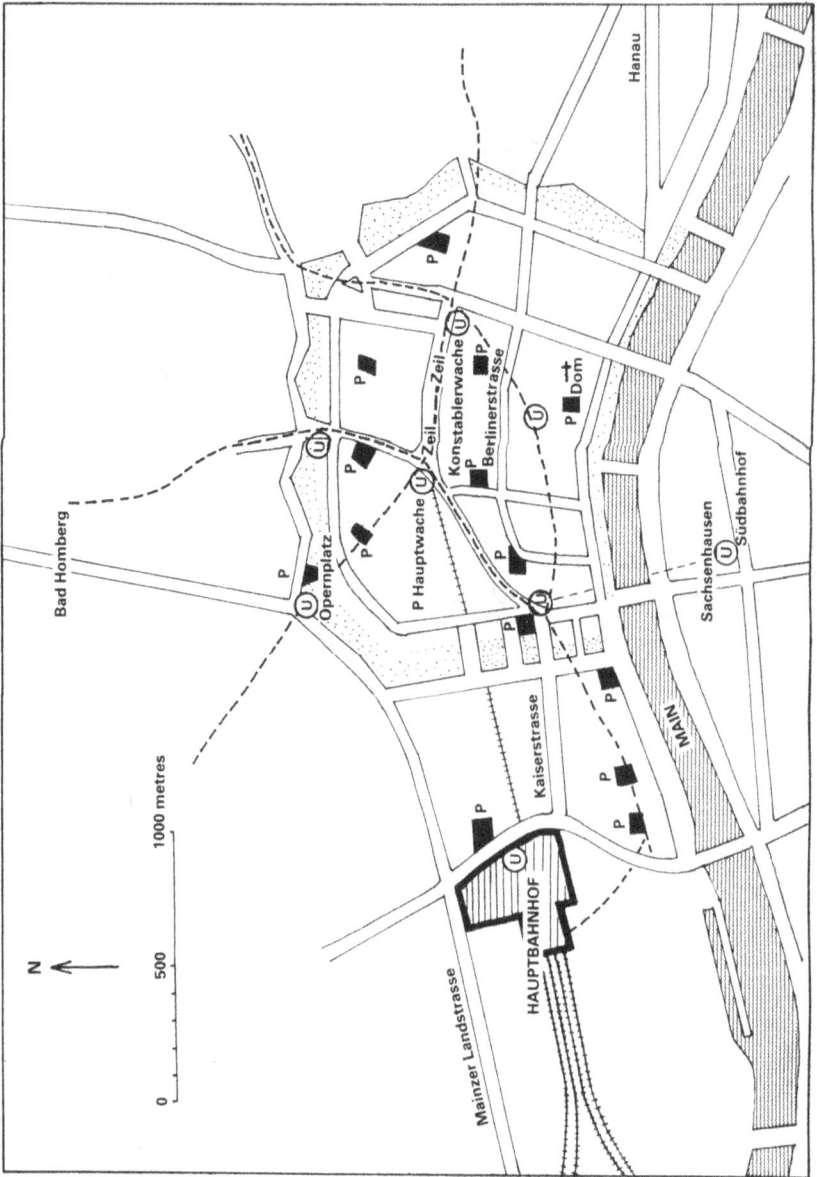

Fig. 7.13 Frankfurt city centre.

Südbahnhof and with the Hauptwache at the centre (Stadt Frankfurt am Main 1976).

Apart from the Hauptwache and Konstablerwache, there are only three other *U-Bahn* or *S-Bahn* stations within the inner ring. Theaterplatz station (Fig. 7.14) is on the southerly extension of the *U-Bahn* lines 1, 2 and 3 from the Hauptwache. Recently this has been extended south of the River Main to the Südbahnhof where it connects to *S-Bahnen* 7 and 8 serving the eastern suburbs and outlying towns. Taunusanlage is an intermediate station between the Hauptwache and the *Hauptbahnhof* on *S-Bahnen* 1–6 and 14. Eschenheimer Tor is on the main northern *U-Bahn* route (Lines 1,2 and 3) leading to the Hauptwache.

Only two bus routes enter the inner ring (actually following the same route here). The *Strassenbahn* acts as an alternative for short distance journeys and 9 of the 15 routes penetrate into the inner ring.

There are plans to replace the *Strassenbahn* and amalgamate it with some sections of the *U-Bahn* to form a *Stadtbahn* (Frankfurt am Main Dezernat Planung 1984). It is envisaged that there will be four cross-city routes extending over a total of 85 km, 39 km from existing tracks.

Fig. 7.14 Theatreplatz, an important changing point between the *Strassenbahn* (eight lines) and the *U-Bahn* (four lines), recently connected by the *U-Bahn* to the *S-Bahn* network at the Südbahnhof 2 km to the south-east.

Fig. 7.15 The Opera House, built in the north-western part of the former fortifications surrounding the inner city.

Fig. 7.16 Two one-way streets, one each side of the former fortifications, act as an inner ring. Much of the city-centre car parking is in multi-storey buildings off these minor streets leading from the inner ring.

Fig. 7.17 Riverside walkways and cycleways line the banks of the River Main in Central Frankfurt.

Conclusions

In the large cities, the *Mineralölsteuer* has allowed heavy capital investment in abundant rapid rail transit to fuel a booming local economy. These West German characteristics are perhaps even more acute in Frankfurt than in Hamburg. The distinction between *U-Bahn* serving the city and *S-Bahn* serving the region are sharper in Frankfurt than in Hamburg. In Frankfurt there are clearly two networks – the *U-Bahn* centred on the Hauptwache and Konstablerwache and the *S-Bahn* centred on the *Hauptbahnhof*, the three stations being connected by an underground public transport axis which must greatly reduce surface traffic.

Both cities rely on the same principle of using the *Hauptbahnhof*, peripheral to the city centre, as the main collecting point with limited surface penetration into the very centre of each city; although in Frankfurt most of the *U-Bahn* passengers pass under the centre. In both cities some of the most dense employment is peripheral to the centre (around Berliner Tor and in City Nord in Hamburg, and in the banking and insurance district to the west and north-west of the centre in Frankfurt) – a policy pursued to reduce the overall amount of traffic within the city centre.

The rapid transit networks have permitted extensive pedestrianization and other environmental improvements. In Frankfurt the *Strassenbahn* has continued to take on the job of what in other cities is done by the buses, with further benefits to the environment.

8 Bremen and Lübeck

Population size and density, degree of concentration of centralized activities (such as jobs, shopping, schools and leisure) and other facilities and policies towards private transport are the main factors which determine the economic viability of urban railways. Together with political will and policy towards subsidy, these are the influences which determine whether a *métro*, tramway or other rail-based network will be built. They therefore determine many of the relationships between town planning and public transport planning due to the fundamental differences between the effects of fixed track networks and buses.

There is, of course, no clearly definable boundary in terms of population, density, concentration, or other relevant characteristics, between those settlements which can support and be supported by a rail network and those which cannot.

At late 1986 prices, a small *métro* of 15 km in length and 20 stations, comparable in scale with those in Marseille, Lyon and Lille, would cost around £18 to £20 million per year to run, including loan charges. Assuming 25% subsidy and a 50:50 split between public and private transport, this would imply journeys typical of at least 500,000 population to be viable, assuming that the *métro* accounted for 30% of public transport journeys. It is, of course, possible that the cheaper forms of railway such as magnetically levitated or suspended railways could make rail-based networks viable in much smaller settlements. So far, however, especially in Britain and France, settlements below the second tier rely heavily on buses for their public transport, supplemented by limited use of what are basically non-local railways for local traffic, at least in Britain.

We now turn to two West German cities which do not have *U-Bahn* or well developed *S-Bahn* serving the city centres. Bremen relies heavily on the *Strassenbahn* and Lübeck, like French and British cities of a quarter of a million population, on buses.

Bremen

Bremen is the smallest of the eleven West German *Länder* or states both in terms of area (404.23 square kilometres) and population (548,000 in December 1982: Freie Hansestadt Bremen 1983 *Flächennutzungsplan*). The population, has, in fact, declined by 60,000 since the peak in 1969. Only 3,600 live in the *Innenstadt* but 43,000 work there (Freie Hansestadt Bremen 1984). The population density within 1 km of the centre is therefore low by German standards, at 5,500 persons per sq. km (in 1981) but the inner ring between 1 and 3 km is more characteristically German at 11,000 persons per sq. km.

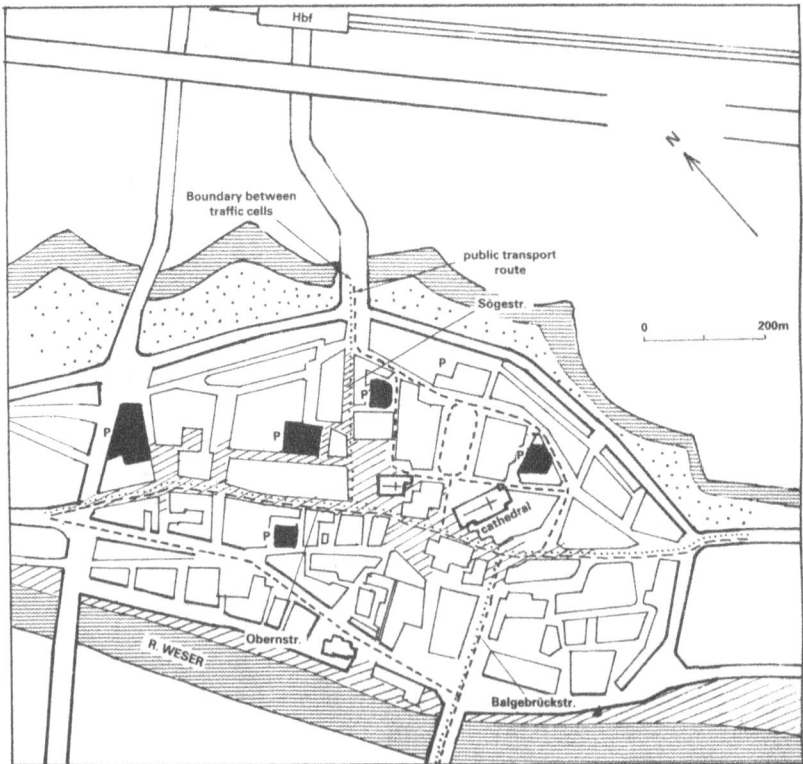

Fig. 8.1 Bremen city centre.

The administrative area is elongated in shape from north-west to south-east along the River Weser with a detached area of port uses, also part of the *Land,* in Bremerhaven, 50 km downstream on the North Sea Coast. Bremen and Bremerhaven are together Germany's second port after Hamburg in terms of goods handled. The administrative area is tightly defined, extending only up to about 10 km from the Weser and mostly less than this. Particularly in the north, it cuts through the urban area of Blumenthal, Vegesack and Binglesum and substantial settlements outside the *Land* look towards Bremen as the Central Place.

The overall structure of Bremen suburbs and neighbouring towns in Lower Saxony is planned around a series of axes extending north, north-west, south and south-east from the centre, based on rapid transit (Hollmann 1977).

Fig. 8.2 The *Hauptbahnhof* in Bremen is a short walk away from the city centre, which relieves congestion.

By West German standards, car ownership is low (3.45 inhabitants per car compared with 2.58 in West Germany as a whole). Car parking in the city centre is restricted with 4,100 off-street places, although there are 19,000 on- and off-street car spaces in the CBD.

The city centre presently lies mainly north-east of the River Weser, extending outside the old fortified semi-circle bordering the river, north eastward to the *Hauptbahnhof*. In the older section of the *Innenstadt,* one third of the street length has been reserved for pedestrians (Freie Hansestadt Bremen,1984). The main up-market shopping street (Sögestrasse), the market area and the Medieval Old City, the Schnoor-Viertel, are pedestrian and Obernstrasse, another major shopping street is pedestrian and *Strassenbahn*. By West German standards, pedestrianization is not extensive, and it is possible that it will be extended in order to stop the present decline in shopping.

The city centre is renowned for its traffic restraint. For transport and land use planning purposes, the city centre has been divided into four cells separated by the main pedestrian streets, Sögestrasse

Fig. 8.3 In Bremen the *Strassenbahn* forms a figure-of-eight in city centre with six branching radials from it. The total length is 57 km.

(shopping) from north-east to south-west, and Obernstrasse (north-west to south-east), and Balgebrückstrasse, one of the six main access roads to the *Innenstadt*. These four cells are contained within peripheral main transport routes which give access to parking, mainly inside the inner ring road at the interface with the mostly pedestrian streets right in the very centre. Road traffic can enter or leave each cell from the ring road but cannot transfer between cells. This has reduced the need for light rapid transit on the same scale as other large West German cities. In fact, Bremen is one of the largest West German cities not to have a well-developed *S-Bahn*, *U-Bahn* or light rapid transit system to service the city centre. The only section of line which may be described as *S-Bahn* is the section between Bremen *Hauptbahnhof* and Bremen Vegesack 18 km north-west of the centre. The only light rapid transit line is a 10 km section in the suburbs.

Public transport penetrates into the pedestrian area. Public transport, and particularly the trams, are important in maintaining the centre. Public transport is used by 48% of those travelling to the

Fig. 8.4 Sögestrasse is one of Bremen's main pedestrian shopping streets. Pedestrian flow increased by 7.6% per year after closure to traffic (Monheim 1980).

city centre (compared with 24.5% of all passenger journeys), 45% travel by car and 7% by bicycle or motorcycle. All six tram lines are cross-city and pass through the city centre. Two-thirds of the 34 bus routes are confined to the suburbs and bring passengers to the tram. Three are cross-city and four others terminate in the city centre. A few also act as feeders to the railway. The trams are particularly significant in giving access to the pedestrian areas and are the only form of passenger transport allowed in several of the main pedestrian streets.

Amongst large West German cities, Bremen is unusual in pursuing city-centre traffic policies so restrictive that a *U-Bahn*, light rapid transit or even a substantial *S-Bahn* has proved to be unnecessary. The trams have been modernized as the backbone of the public transport network and are particularly significant in giving access

Fig. 8.5 Trams come into the historic centre of Bremen with little damage to the environment.

N
↑

bus station and
railway station

Holstentor platz

Holsten Staße

Trave

Stadt

Kanal

Trave

Marien-
kirche

Markt

cathedral

0   200   600m

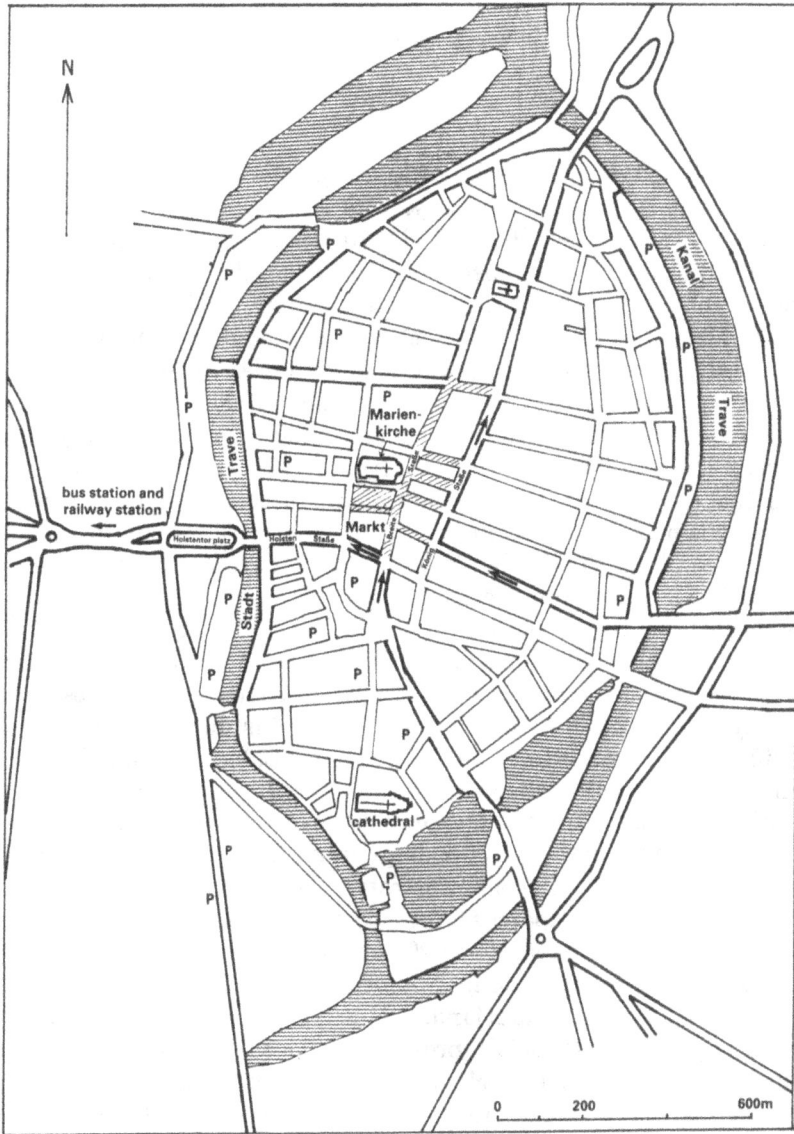

Fig. 8.6 Lübeck *Innenstadt*.

right into the pedestrian area. The bicycle is also an important means of local transport, and there is some evidence that this substitutes and provides competition for public transport.

Lübeck

The Hanseatic Port of Lübeck is situated on the Baltic coast adjoining the boundary with East Germany. The population has declined from 239,300 in 1970 to 215,800 in 1985. Although the population of the central area, too, has declined from 18,500 to 15,700, it is still quite considerable, and in fact increased slightly after 1980.

The settlement is of a rather elongated shape from north-east to south-west. The *Innenstadt* is situated on an island of 100 hectares in the River Trave, surrounded by several physically separate suburbs, divided by waterways and man-made barriers such as railways and *Autobahnen* and by non-residential uses. The *Innenstadt* is 15 km from the open sea to the north-east, but linked to it by the River Trave. Main lines of communication link the city to Schlutrup, Kucknitz and, on the Baltic, Travemünde, a significant seaside resort and port. All are within the administrative city of Lübeck.

Being on an island, access to the *Innenstadt* is restricted to the eleven bridges. Both the bus station and the main railway station are 300 m west of the *Innenstadt* which relieves some of the congestion of the city centre. Building density of the island is, however, high, with significant residential uses, as well as being Lübeck's main shopping, commercial and cultural centre.

Public transport in Lübeck is provided by the *Deutsche Bundesbahn* which operates five lines leading from Travemünde and other destinations outside the administrative area to the *Hauptbahnhof*, and by six bus companies. One company run by the city, the *Stadtwerke Lübeck,* operates 90% of the bus routes. Of the other companies, the *Bundesbahn, Autokraft, Post* and *Dahmetal* provide services in the hinterland, and the *Lübeck-Travemünde-Verkehrsgesellschaft (LGV)* provides services in Travemünde and from Travemünde to Lübeck. In Lübeck as a whole, 550,000 daily journeys are made (Hansestadt Lübeck, der Senat der, 1978) of which 385,000 are by private car or cycle and 147,000 are by bus, 9,400 by rail and 8,600 by bus and rail.

Lübeck is quite a highly centralized city and, as a result, the level of mobility at 1.89 journeys per person per day is high in comparison with the Federal Republic as a whole (1.7). Of the 1.89 journeys, 1.3 are by private and 0.59 by public transport. The public transport rate is high compared with similar cities, largely due to the constrained nature of the inner city which restricts the private car even more than public transport.

With 310 private cars per thousand inhabitants, car ownership is less than the national average (350). As in many German cities, the bicycle is a serious competitor to local public transport. In fact about 60% of all Germans and 85% of households own at least one cycle. Although only 12% of journeys in Lübeck are by cycle, it may contribute to any consumer resistance to fare rises on public transport. There is quite a comprehensive network of cycleways with 111 km of separate routes and a further 140 km of secondary routes separated from the traffic or footpath (Hansestadt Lübeck, der Senat der, 1985). The concentration of commerce, shopping, jobs and other activities in the city centre, the shape of the city, elongated from north-east to south-west and the relatively great distance between residential areas and industry all result in quite long average journeys: 8.3 km for all journeys, 9.0 km for those by private transport and 6.8 km for public transport.

Fig. 8.7 In Lübeck, access to the *Innenstadt* is restricted to the eleven bridges.

In parts of the city, particularly Travemünde, weekend traffic is important in planning needs for transport infrastructure. Some Lübeck roads have more traffic on Sunday afternoons than at weekday peak periods. Travemünde receives many visitors from Lübeck, Hamburg, Niedersachsen and Schleswig-Holstein, and in parts, weekend traffic is twice the weekday peak.

Transport planning in the *Innenstadt*

Because of its restricted island site, the *Innenstadt* has become densely developed over the centuries. Shopping and commercial uses appropriate to a city centre extend along two parallel north-south thoroughfares, Königstrasse and Breite Strasse and westwards along Holstenstrasse in the direction of the *Hauptbahnhof* and *ZOB* (bus station). The *Innenstadt* is an important cultural centre with seven churches including the Cathedral and the Marienkirche, renowned for centuries for its music. Many of the buildings are of architectural or historic interest. Nearly all the streets in the *Innenstadt* have some such buildings. Apart from the main shopping streets, Königstrasse, Breite Strasse and Holstenstrasse and a few minor streets near the Marienkirche, many of the frontages are composed mainly of classified buildings (Hansestadt Lübeck, der Senat der, 1984).

Fig. 8.8 Travemünde is a significant port and seaside resort.

Planning and transport policies for the *Innenstadt* have been to maintain and strengthen three main functions: as a cultural and historic centre, as a residential area for various social groups and as a shopping centre and Central Place (Hansestadt Lübeck, der Senat der, 1973). These impose differing and in some cases conflicting demands on public transport. The large number of fine historic buildings, many set in narrow streets, imposes severe restrictions on a city dependent on buses for public transport. In the *Innenstadt* buses are restricted to a few streets.

Fig. 8.9 Travemünde.

The planning of both public and private transport has been guided by the principle of stopping on the edge of the *Innenstadt* with limited penetration into it. A series of tangential and ring roads outside the *Innenstadt* are intended to act as by-passes, filtering off through traffic. The railway is entirely outside, the *Hauptbahnhof* being about 300 m to the west. The *ZOB* for local and long distance bus and coach services is next to the *Hauptbahnhof*. Heavy bus flows follow a limited number of streets within the *Innenstadt*. Stops are grouped together to help interchange.

There is parking for 8,900 vehicles, mostly on former industrial and port land on the periphery of the island. Some multi-storey car parks have been built closer to the historic centre but the tariffs are higher at DM1.50 per hour (August 1985). Apart from peripheral main roads, nearly all those which penetrate into the *Innenstadt* are one-way.

Transport problems are addressed at various levels. The *GVP (Generalverkehrsplan)* (Hansestadt Lübeck, der Senat der 1978) applies to the whole conurbation. The *Innenstadt* is planned as a whole in terms of addressing problems such as parking, diverting through traffic, ensuring access to commercial and cultural activities and the conservation of historic buildings and the environment. Problems and alternative solutions are analyzed by topic and by transport route (Hansestadt Lübeck, der Senat der, 1982).

More locally, the *Innenstadt* is conceived as comprising six districts,

Fig. 8.10 Pedestrianization is limited to Breite Strasse and a few minor adjoining streets to the east and the area around the Marienkirche to the west.

each with its particular problems and solutions, the boundaries being mostly formed by the River Trave and the principal roads. For example, the south-western district around the Cathedral is characterized by problems of cars causing disturbance to local residents whilst touring the streets in search of restaurants and entertainment facilities as well as the more widely applicable problems of remaining through-traffic and drivers searching for parking. Physical measures and scaled parking charges have been used to address these problems (Hansestadt Lübeck, der Senat der, 1982). In an earlier model (Hansestadt Lübeck, der Senat der, 1973), the city centre had been viewed as comprising four zones divided by Breite Strasse and Königstrasse, each planned in relation to bridges giving access to the island. Access to a part of the *Innenstadt* on the opposite side of the centre is intended to be via main roads outside the Island rather than by crossing the *Innenstadt*. In actual fact, ten alternative variations

Fig. 8.11 Königstrasse, main shopping street and bus interchange.

on the themes of historic centre, Central Place and residential area were formulated (Hansestadt Lübeck, der Senat der, 1973). This involved some increase in shopping, business and other central place uses at the expense of mixed uses *(Baunutzungsverordnung)*. There was also to be some increase in the residential-only area to replace mixed residential areas with other uses.

Table 8.1 Lübeck: existing and preferred land uses.

	Existing %	Preferred %
Kerngebiet (central area uses)	12.0	16.0
Mischgebiet (mixed land uses)	19.0	14.0
Allgemeines Wohngebiet (residential and ancillary uses)	18.0	16.0
Reines Wohngebiet (residential areas)	16.0	21.0

The main access to the commercial centre would concentrate further on the western access to the *Hauptbahnhof* and bus station, with three further access routes north, east and south.

Fig. 8.12 Pedestrianization is not very extensive, even in the main shopping area of Lübeck, Königstrasse.

Conclusions

Bremen and Lübeck are examples of traffic restraint being used as an alternative to heavy investment in local rail transport. In both, the principle of inner ring and cells has been used to control traffic entering the city centre. *Strassenbahn* in Bremen and buses in Lübeck operate on a few streets to bring passengers into the very centre. In each city, railways are limited to the periphery of the centre and there the *Hauptbahnhof* acts as a transport node. Pedestrianization is limited by West German standards, perhaps reflecting the lack of underground railways.

9 Paris and London

Some of the relationships between town planning and public transport are very much dependent on the size of settlement. However difficult to define, there are minimum settlement sizes for which a *métro* would be viable. There are similarly ill-defined maximum settlement sizes for which walking can be expected to be the main form of transport, and therefore for which public transport will be necessary for large numbers of people.

For some of these definitions of settlement size, population will be a reliable indicator. In many cases, other characteristics will be significant as well. Size of city centre in terms of number of jobs, attractiveness of shopping, leisure and other facilities may be more significant than population in indicating journey patterns and the form of transport that is likely to be economic. In Central Places generally, the relationships between town planning and public transport will be different from those in settlements of equal population but less centrality.

Other factors as well as population and centrality which affect the relationships between town planning and public transport include building density, housing, shopping, employment and other uses causing journeys. All of these factors – population size, centrality, density and dispersal – are not likely to be independent of each other. The first three in particular tend to vary in proportion to each other.

These are the criteria which relate the economics of public transport to urban development. In some cases, the choice of transport infrastructure is at least as much influenced by political choice. The fundamental difference between rail based public transport in the large West German cities and road based transport in French and British cities is certainly not explicable in terms of differences in physical forms and sizes of settlements. When we look at smaller towns in the three countries, however, the contrasts are not so striking.

Birmingham, Lyon, Marseille, Frankfurt and Hamburg are all second-tier cities. With the annexation of part of the West German primate city, Berlin, and the isolation of the remainder of it from West Germany and the transfer of some capital functions to Bonn, the other capital functions have become distributed amongst the second tier cities. Birmingham is perhaps more under the influence of London, only 180 km distant, than Lyon is influenced by Paris, 430 km to the north, and certainly more than Hamburg is influenced by Berlin or Bonn. Despite these differences, all three cities share a lot in common in terms of size, centrality and other factors likely to influence planning and public transport. They show only a limited range of the ways in which planning and public transport are intertwined. We will therefore now look briefly at some settlements very different in character in order to expand our survey of planning and public transport.

Paris

The larger the city, the greater is the economic incentive for rail transport. London and Paris each support very well developed underground and suburban rail networks (Table 9.1).

Table 9.1 Public transport in London and Paris.

	Total length (km)	Stations	Passengers p.a. (millions)
London underground	397	279	563 (1983)
Paris *métro*	198	365	1,166 (1986)
London suburban trains	770	297	469 (1983)
*RER**	103	62	291 (1986)

	Number	Length of route (km)	Passengers p.a. (millions)
London buses	4,980	5,839	1,160 (1983)
Paris buses	3,996	2,910	795 (1986)

Réseau Express Régional operated by *RATP*. It is planned to extend this to several SNCF routes by the mid-1980s.

Sources: London Transport 1984; *RATP* 1986.

In terms of transport infrastructure, the large West German cities more closely resemble a scaled-down version of London or Paris than their (larger) British counterparts such as Birmingham or Manchester. It might be expected that Frankfurt, Hamburg or other large German cities would not be able to compare with London or Paris in their financial performance, simply because they do not have such large populations. Nevertheless, their subsidies are no larger than in Paris.

Parisian local transport is distinctly more local than that in London. Route length is less, and at least on the *métro*, stops are more closely spaced. The Paris *métro* is largely confined to the *Cité de Paris,* whereas the London Underground extends almost as far as the Greater London boundary in the north and west and, in the cases of the Metropolitan and Central lines, well outside. The *RER* has been

Table 9.2 Paris *Métro.*

Line	From	To	Stations	Main line stations
1	Pont de Neuilly	Château de Vincennes	22	Lyon
2	Porte Dauphine	Nation	25	
3	Pont de Levallois-Bécon	Gallieni	24	St Lazare
4	Porte de Clignancourt	Porte d'Orléans	25	Nord, Est, Montparnasse
5	Eglise de Pantin	Place d'Italie	20	Austerlitz, Nord, Est
6	Charles de Gaulle/Etoile	Nation	28	Montparnasse
7	Fort-D'Aubervilliers	Mairie d'Ivry	33	Est
7b	Louis Blanc	Pré St Gervais	8	
8	Balard	Créteil-Prefecture		
9	Pont de Sèvres	Mairie de Montreuil	37	
10	Boulogne-J Jaures	Gare d'Austerlitz	21	Austerlitz
11	Chatelêt les Halles	Mairie des Lilas	13	
12	Mairie d'Issy	Porte de la Chapelle	28	St Lazare, Montparnasse
13	Gabriel-Peri Saint-Denis-Basilique	Châtillon-Montrouge		St Lazare, Montparnasse

Notes
1 Lines 1 and 4 are close to the north west-south east and north-south axes of the City's communications respectively.
2 Lines 2 and 6 function as a circle along the line of the exterior boulevards. There is only one main line station on them and no circular route further out from the City.
3 Line 8 follows most of the northern section of the interior boulevards.
4 There is no direct route between Gare du Nord and Gare de Lyon.

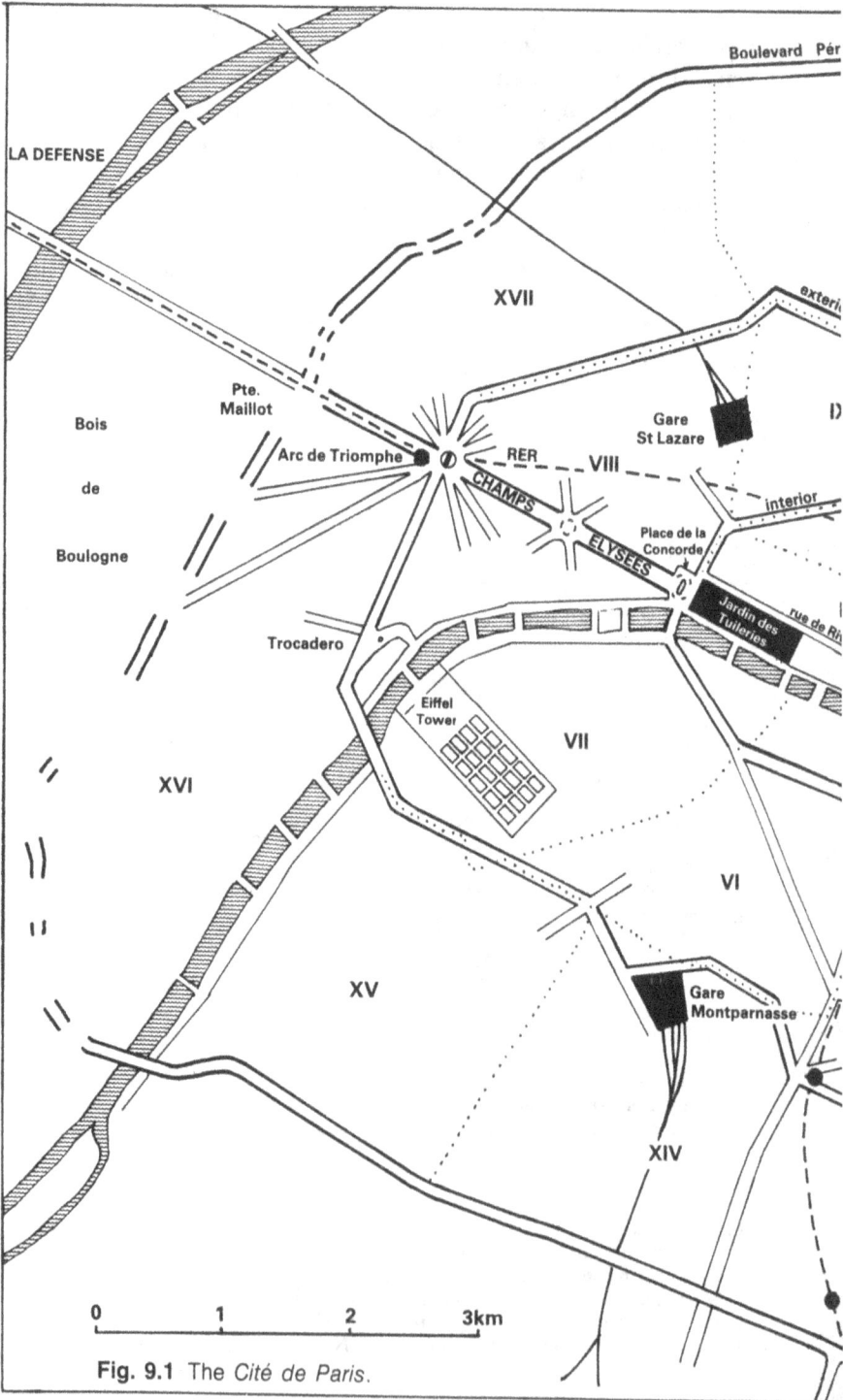

Fig. 9.1 The Cité de Paris.

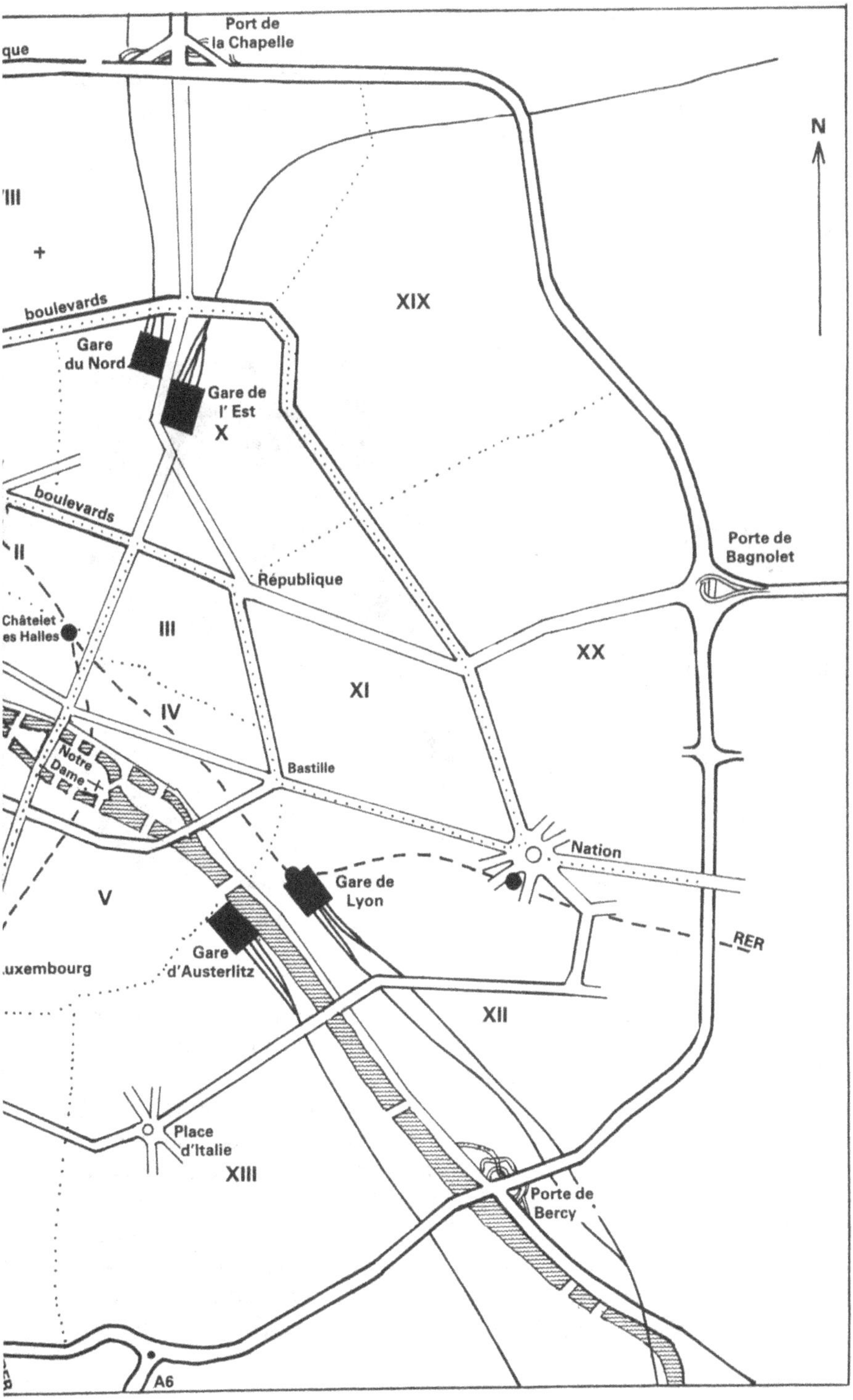

Port de
la Chapelle
que
N
'III
XIX
boulevards
Gare
du Nord
Gare de
l' Est
X
boulevards
Porte de
Bagnolet
II
République
Châtelet
es Halles
III
XX
XI
IV
Notre
Dame
Bastille
Nation
V
Gare de
Lyon
RER
Gare
d'Austerlitz
uxembourg
XII
Place
d'Italie
XIII
Porte de
Bercy
A6

developed since the mid-1960s to extend the rapid transit network to the middle and further suburbs which are in the main out of the reach of the *métro*. Long-distance commuting has never been as readily accepted in France as in Britain. Recently the RATP has been turning attention to districts of medium density as well as the higher densities served by the *métro* and *RER*, and a light rail line is proposed in the north-east of Paris (International Railway Journal, November 1985, pp. 45–8).

Fig. 9.2 The Champs Elysées, tourist mecca and part of the north-west/south-east axis through central Paris.

Perhaps more significant, Paris is distinctly smaller in area and more densely developed than is London. The 1,146 sq. km of the RATP area had a population of 7,161,000 in 1982 (Régie Autonome des Transports Parisiens 1983d), whilst at the time of the 1981 Census the Greater London Council area of 1,579 sq. km had a population of 6,713,165. If we look at the inner areas the contrast is even more striking. In 1981 the 13 inner London Boroughs and the City of London had a population of 2,497,978 within 321 sq. km at a density of 7,790 persons per sq. km. The population in 1971 was 3,031,936 and in 1961 3,492,879. The *Cité de Paris* had a population of 2,176,000 within 105 sq. km in 1982 (density 20,724 persons per sq. km).

The 1965 *Schéma directeur d'aménagement et d'urbanisme de la Région de Paris* set out main directions of urban growth which have guided the overall planning of the capital since then. Because of the very high density, the developed area was expected to increase from 1,200 to 2,300 sq. km. There were two major south-east to north-west axes. The northern one stretches 75 km through Marne-la-Vallée and Cergy-Pontoise; the southern axis extends 90 km through Melun-Senart, Evry and St Quentin-des-Yvelines. Six large suburban centres were to be created 10–15 km from the centre at Créteil, Rungis, Versailles, la Défense-Nanterre, St Denis and Bobigny. These serve populations of 300,000 to 1 million, a recognition of the over-dependence on central Paris and the absence of any centres with more than purely local services in the suburbs. The *RER* was proposed on a south-east to north-west axis through central Paris from Boissy-St Leger to St Germain with a north-south line connecting at Les Halles in central Paris.

Since 7th January 1959, the overall organization, economic management and technical coordination of public transport in Greater Paris has been the responsibility of the *STP (Syndicat des Transports Parisiens)*. This covers an area of 4,007 square kilometres including the *Cité of Paris*, the three adjacent *départements* of Val-de-Marne, Hauts-de-Seine and Seine-Saint-Denis and the more urban *communes* adjacent *départements* (RATP 1981a). At the head of the *STP* is a council of 20 members. The State is heavily represented with ten members, the remainder divided between the *Cité of Paris* (5) and the other *départements* (5).

Unlike the *STCRL* in Lyon (chapter 2 above), the *STP* does not have a single executive body responsible for coordination of services (*STCL* in Lyon) or an organisation responsible for research as is *SEMALY* in Lyon. Services are, in fact, run by three organisations, and the *STP* decides the distribution of these services between them:

1 *RATP (Régie Autonome des Transports Parisiens)*, responsible for services within the central part of the region, covering an area of 1,146 sq km and with a population of 7,161,000 in 1982.

2 *SNCF (Société Nationale des Chemins de Fer Français)*, which serves suburbs and towns up to about 50 km from Paris to the large railway termini on the fringes of the city centre.

3 *APTR (Association Professionelle des Transports Routiers)* which represents about 60 private bus and coach firms operating services mostly in the parts of the Ile de France outside the area of the *RATP*. There are, however, some private services in the new towns of Cergy-Pontoise, Evry and Saint-Quentin-en-Yvelines operating under charter from the *RATP (RATP 1983b)*.

The functions of the *STP* include both policy-making and project approval. It is responsible for dividing between its members (State, *Cité de Paris* and *départements)* the cost of deficits, including the receipts lost as a result of reduced fares for period tickets and the compensation for not increasing fares (*indemnité compensatrice: RATP* 1981b). In 1983 the State bore 70% of this burden, the *Cité de Paris* 22% and the *départements* 8%. Conversely, the *STP* distributes the income from *versement transport* between the *RATP*, the *SNCF* and the *APTR*. The *STP* decides routes, frequency of services, building of new stations or closure of existing ones. It approves investment projects and oversees the accounts of the *RATP* and *SNCF*. In addition it determines fares to ensure coordination between services and overall viability. If the Minister of Transport opposes a fare increase proposed by the *STP*, compensation may be payable to the *STP (RATP* 1981b).

Fig. 9.3 Traffic flow is not so much of a problem on the boulevards as at the *ronds points* or junctions.

The *RATP* was formed on 1st January 1949 as a public authority taking over the public transport networks previously belonging to the *Compagnie du Chemin de Fer Métropolitain de Paris* and the *Société des Transports en Commun de la Région Parisienne*. It runs the *métro, RER* and the buses. The *RATP Plan d'Entreprise (RATP* 1984b) is an important planning document setting out the relationships to the National Plan, analysis of factors affecting the provision of public transport such as demographic changes, commercial, research and service strategies, management, industrial relations, plans for *métro, RER* and buses – in fact all the functions of the undertaking. It is produced each year as a five-year rolling programme.

The *Plan d'Entreprise* is both that of a commercial firm and that of a public service (*RATP* 1983e). For example, there is the desire to extend the public transport network further into the suburbs to capture more of the journeys entirely confined to the suburbs. On the one hand, this provides a public service by reducing the inequality of service between city and suburbs. On the other hand, only 15.5% of suburb-to-suburb journeys are by public transport (compared with 59.5% from Paris to suburb and 69% Paris to Paris: all figures for 1983), and this is thought to be the kind of journey with greatest potential for capture by public transport.

Fig. 9.4 The Jardin des Tuileries.

The *RATP* describes itself as a public undertaking and is conscious of the duty to run routes in deficit when public interest requires and to put into practice policies which only indirectly serve its interests. For example, there have been considerable attempts to improve the environment and to put a 'human face' on the *métro* – a policy widely supported by public opinion but which will not increase the resources of the undertaking except in so far as there might be some increase in traffic.

The *Plan d'Entreprise* reflects central government policy in several ways. Much attention is given to the decentralization of decision making within the undertaking. The intention to improve suburban transport is not only a reflection of commercial criteria but is also in line with the 'right to transport for all' in the *Loi d'orientation des transport intérieurs* and the Ninth National Plan. Also, the policy towards stabilization of prices in the Ninth National Plan is clearly reflected in the *RATP* fares policy, which is to the effect that rises in fares should be in line with rises in disposable income.

The question is raised as to the possibilities of a wider range of financial support for public transport. It is pointed out that shops, developers and motorists, for example, also benefit from the effects of public transport yet do not contribute to it directly (except for those who employ ten or more people by way of *versement transport*).

Other financial policies, or perhaps more accurately recommendations, are that extensions to the networks should be met from public funds, renewal of rolling stock and other equipment should be self-financed and investment in modernization and improvement leading to an increase in benefits or a decrease in costs should be self-financed or covered by loans.

The *Plan d'Entreprise* covers an area and time period where changes in land use will be small in proportion to those already in existence. The *Schéma Directeur d'Aménagement et d'Urbanisme* is mentioned only in passing, but past plans have fundamentally influenced the need for public transport. The introduction of polycentric growth in the 1965 *SDAU*, for example, with restructuring poles in the suburbs and new towns along preferred growth axes, has had a profound effect on transport requirements.

Both *Plan d'Entreprise* and *SDAU* show a common interest in demographic and other changes which are leading to an outward migration of the market for public transport. As part of the measures to combat this and to reduce the difference in servicing levels between the centre and the periphery of the conurbation, a series of public transport networks is proposed based on three rings: *petite ceinture* (inner ring), *rocade intérieure* (inner circle), 2 km from the Paris boundary, and the *rocade extérieure* (outer circle), 5 km from the Paris boundary. *SNCF*, *métro* and *RER* will form the radials, linked together by busways.

London

Whereas the population of Greater Paris has been increasing in recent decades and by an average of 0.4% between 1975 and 1982 (Cahiers Français 1985a), that of the area of the former Greater London Council has declined from 7,452,346 in 1971 to 6,713,165 in 1981.

The M25 ring around London was completed in 1985 and is the only true orbital route. It is mostly outside the GLC area, and is 192 km in length. It is, of course, intended to relieve the pressure of traffic in London and to provide a more rapid route across the built-up area. The latter it certainly does, but it has been accused of accentuating the decentralization from London to the periphery, particularly to the already prosperous west, already under pressure for development. It will do little for Central London and may even be detrimental economically (Damesick, Lichfield and Simmons 1986). The south circular road and the inner ring are really just general purpose roads including high streets and residential streets strung together by signposts. However, large-scale road construction within London to reduce congestion seems unlikely in the immediate future. The Greater London Development Plan, finally approved in 1976, was left with no major new road proposals after the deletion of the proposed new orbital routes and Ringway One, somewhat equivalent to the *Boulevard Périphérique* in Paris. The main effort has been to the improved management of the existing road network in the forms of control of car parking (although the number of traffic wardens is less than one third that estimated to be necessary by the Metropolitan Police), reduction of through-traffic and priority for public transport.

In London, the red buses and underground were operated by the London Transport Executive, policy and financial control being exerted by the GLC until 29th June 1984 when the Executive was replaced by London Regional Transport responsible to the Department of Transport. Thus local transport passed away from the control of local government to central government.

Despite this, up to two-thirds of subsidy can be levied by central government on the local authority (London Regional Transport Act 1984). London Regional Transport owns two separate companies, London Buses Ltd., and London Underground Ltd. which run their respective parts of the transport network.

Between 1982 and 1985/86 there was a considerable increase in usage of public transport. Passenger journeys on the buses increased from 1,040 million to 1,146 million and on the underground from 498 million to 762 million. Miles per employee rose by 12% on the buses and by 8% on the underground. The proportion of running costs

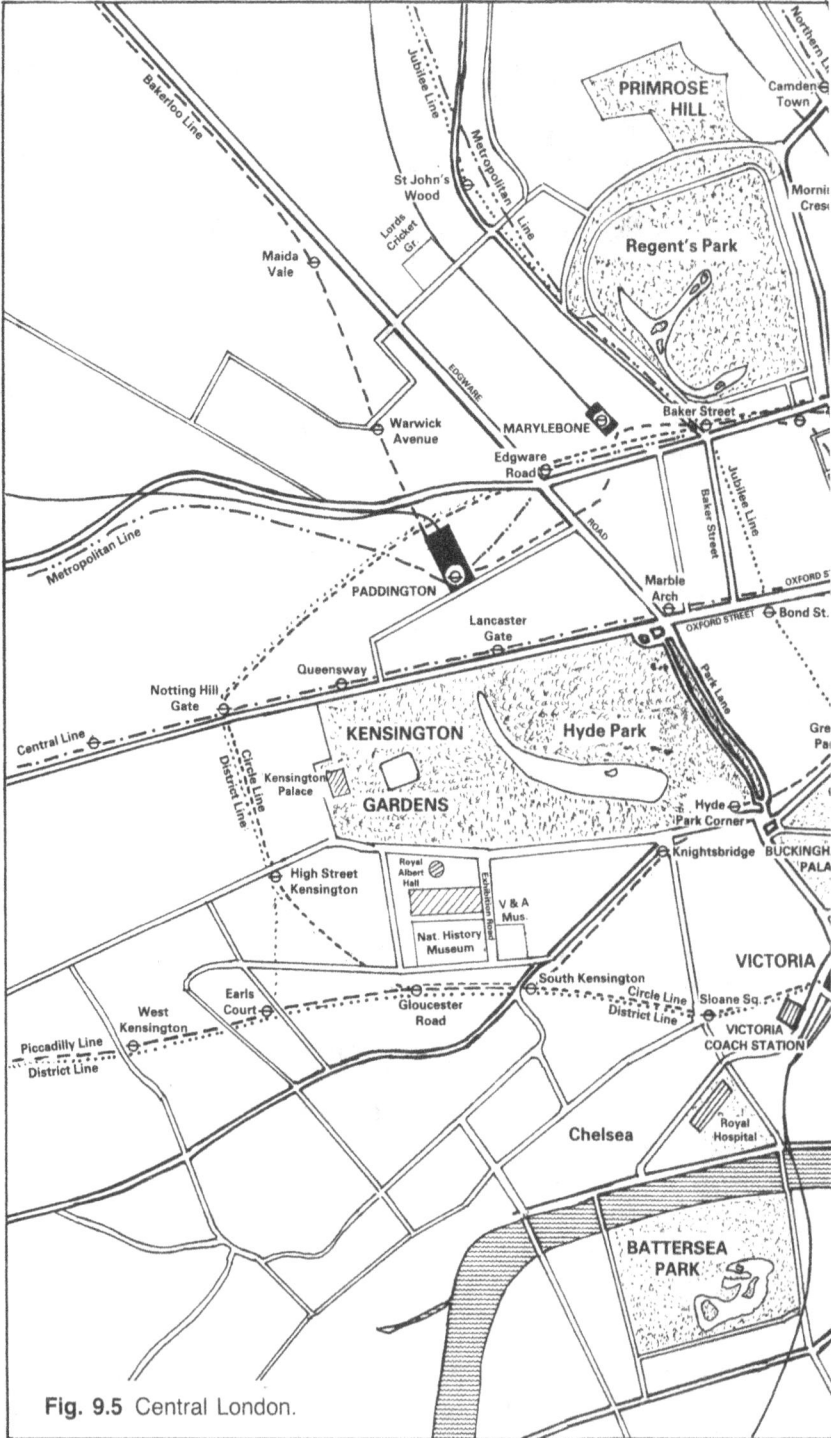

Fig. 9.5 Central London.

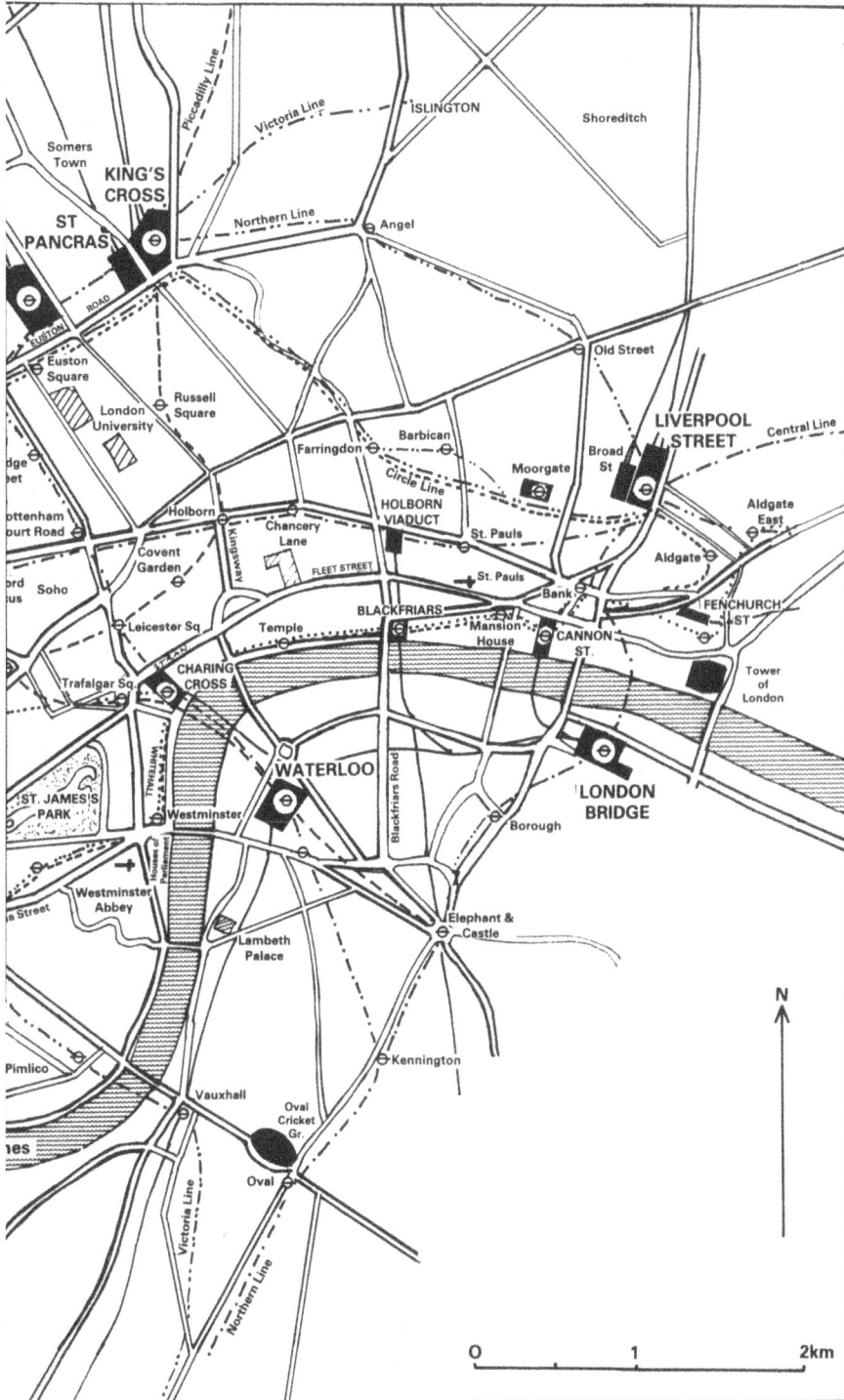

covered by income rose from 62% to 67% on the buses and from 81% to 82% on the underground.

Suburban trains have continued to be operated by British Rail. In addition, there are London Country Buses, Green Line and National Express operated by the National Bus Company, mostly providing services on the outer fringes of the city, sub-regional and regional services.

Until the creation of London Regional Transport, the London Transport Executive had to approve all licences to operate stage carriage services. Most of them, except some coaches for commuters, works or schools, receive some financial support. In the year following June 1984, the number of licensed services not operated by LRT was halved, most of those remaining being provided by London Country Buses and private-sector companies.

The British Railways Board continues to operate train services in Greater London and further afield. This is, of course, a fundamental difference from the six metropolitan transport authorities in Britain which were responsible for railway services within their boundaries. The difference was due to the size and attractive power of London: commuter services extend well outside the Greater London boundary and so a transport authority for London could not represent them. Also, British Rail commuter services are part of a national network of routes, and so funding them separately would be very difficult. In fact, nearly half of British Rail commuters live outside the GLC, whilst underground commuters live mainly within it (Foulkes, 1983).

Use of the underground is particularly high in the west (Hammersmith, Kensington), north (Camden, Harrow, Haringey) and parts of the east (Newham, Redbridge). Bus usage is highest in the east (Hackney, Tower Hamlets) and some inner southern boroughs (Lambeth, Southwark). The suburban railways are even more concentrated in the south (Bromley, Lewisham, Bexley, Greenwich, Sutton, Croydon, Kingston, Merton, Richmond) plus Havering:

Table 9.3 Mode of travel to work in London (%).

	Underground	Bus	BR	Car	Other
Inner London (excluding the City)	17.48	23.16	7.60	24.92	26.58
Outer London (19 boroughs)	9.94	13.03	12.33	43.68	21.02

Source: Census 1981, County Report for Greater London, Part 2, p. 44.

There has been revived usage of public transport recently after several years of decline: (Table 9.4).

Table 9.4 Passengers arriving in Central London 7am to 10am daily (thousands).

	Underground	Bus	Rail	Private	Total
1973	495	144	435	188	1 262
1983	448	97	383	211	1 139
1985/6	516	89	401	212	1 086

Source: London Transport Annual Reports 1974 to 1986.

Fig. 9.6 At Victoria, inter-city and suburban trains, underground and local buses all come together. Green Line sub-regional buses, the longer-distance National Coaches and services to Heathrow and Gatwick Airports run from stations nearby.

Table 9.5 The London Underground.

Line	From/To	Stations (km)	Main line stations	Other inter-changes with BR
Bakerloo	Elephant & Castle to Queen's Park with peak hour extensions to Harrow and Wealdston	16(11)	Paddington Waterloo	Willesden Jct. Marylebone Harrow & Wealdstone Queen's Park Elephant & Castle
Central	West Ruislip & Ealing Broadway to Hainhault or	51(84)	Liverpool St	Greenford Stratford West Ruislip

Table 9.5 *Continued*

Line	From/To	Stations (km)	Main line stations	Other interchanges with BR
	Epping; peak hour shuttle service from Epping to Ongar			South Ruislip Eailing Broadway Woodford
District	Ealing Broadway to Upminster with branches to Richmond, Wimbledon, Edgeware Road and Olympia	66(64)	Victoria Paddington	Ealing Broadway Richmond Gunnersbury Wimbledon Charring Cross Blackfriars Cannon Street Barking West Ham Upminster
Metropolitan	Baker St to Amersham with branches to Chesham, Watford and Uxbridge	34(67)		Amersham Chalfont & Lat. Moor Park Harrow-on-the-Hill
	Hammersmith to Whitechapel peak-hour extension to Barking	19(14)	Paddington King's Cross Liverpool St	Westbourne Gr. Farringdon
	Whitechapel to New Cross or New Cross Gate	7(7)		New Cross New Cross Gate
Northern	Morden to Edgeware, Mill Hill East or High Barnet via Bank or Charing Cross	49(58)	Euston King's Cross Waterloo London Bridge	Kentish Town Finsbury Park Strand Charing Cross Balham Elephant & Castle
Piccadilly	Cockfosters to Heathrow Airport Uxbridge with peak hour shuttle from Holborn to Aldwych	52(70)	King's Cross	Finsbury Park
Circle	linking the main railway terminal north of the Thames	27(21)	Paddington Euston King's Cross Liverpool St Victoria	Cannon Street Blackfriars Charing Cross
Victoria	Walthamstow Central to Brixton	16(23)	Victoria Euston	Vauxhall Tottenham Hale Walthamstow Central
Jubilee	Stanmore to Charing Cross	17(23)	West	Hampstead Charing Cross

Notes

1 The underground is mostly north of the Thames. Only the Northern line extends more than a few stations south of the river.

2 The Circle Line links all the main line stations north of the Thames and encloses an area close to the Census definition of Central London, except for the area between Lambeth Palace and London Bridge south of the river. Euston and Victoria were directly linked only as late as 1969 (by the Victoria Line).

3 The Circle Line acts as an interchange – it shares a route in common with the Metropolitan Line in the north and east and the District Line in the south and west. All the stations on it are also on at least one other line and all the London Underground lines are located within it or in common with it. There are 50 stations on or within the Circle Line.

4 All the Underground lines serve the central area, and all except the Jubilee and Circle are cross-city extending to the outer suburbs at both ends.

Fig. 9.7 Despite their decline by 20% between 1971 and 1980, buses still account for half the public transport journeys in Greater London. Buses are particularly well used for journeys between inner and central London.

The London underground (Table 9.5) began to take shape in the 1860s with the opening of the Metropolitan Railway between Paddington and Farringdon in 1863. In 1868 the sections from High Street Kensington and Gloucester Road to Westminster and in 1900 the Central London Railway from Shepherd's Bush to Bank were opened. Extensions to suburban railways took place to Richmond (1887), Wimbledon (1889), Chesham (1889) and Uxbridge (1904). By 1910 about half of the present day underground network was completed and apart from the Victoria and Jubilee Lines the network in the Central Area was substantially complete. Much of the suburban development however, took place in the inter-War years, to Edgware (Northern Line) in 1924, Watford (Metropolitan) 1925, Morden (Northern) 1926, Upminster (District) 1932 and Cockfosters (Piccadilly) 1933 and the Northern Line section to High Barnet was opened in 1940. The Central Line was extended in the late 1940s to Greenford (1947), West Ruislip (1948) and to Leytonstone (1947), Hainhault (1948) and Epping (1949). In Central London, the Victoria Line began operating from Victoria to Walthamstow Central in 1969 and to Brixton in 1971. The Jubilee Line was opened in 1979. A concise history of the development of the underground is contained in London Underground Limited (1986).

Table 9.6 Main British Rail commuter terminals.

Terminal	Areas served
Charing Cross	South-eastern suburbs
Cannon Street	
Holborn Viaduct	South-eastern and southern suburbs
Blackfriars	
Victoria	Southern and south-western suburbs
London Bridge	
Waterloo	South-eastern and south-western suburbs
Waterloo and city Railway	connecting Waterloo with Bank via the underground
Marylebone	North-western suburbs
Paddington	Western suburbs
North London line	North Woolwich to Richmond where it connects with the Southern Region, via north London inner suburbs
Fenchurch Street	Eastern suburbs north of the Thames
Kings Cross	North London services
Liverpool Street	

The Docklands Light Railway is due to open in 1987 and will connect Stratford and the Isle of Dogs with Tower Hill via Poplar and Limehouse. There will be 12 km of track, much of it using disused railway routes, and 16 stations.

Central London

The central area of London as defined in the Census extends over 26.9 sq. km from Paddington in the west to the Tower of London in the east, and from Somers Town in the north to Pimlico in the south. Apart from the area between Lambeth Palace and London Bridge south of the river, it comprises mainly the City of London and the West End north of the river. It includes the City, most of the London Borough of Westminster and parts of the London Boroughs of Camden, Islington, Southwark and Lambeth. The central area road pattern still owes a lot to the medieval layout in the City and the estates in what is now the West End. The scope for environmental management and pedestrianization is limited by lack of alternative routes on which to divert the traffic displaced. It has a population of around 200,000 and 1 million jobs compared with 230,010 population and 1,241,000 jobs in 1971.

Central London is still characterized as being an extremely attractive employment, tourist, cultural and educational centre with entertainment and shopping in the West End. The Greater London Development Plan (Greater London Council, 1976) encouraged the development of strategic centres, as did the 1965 *Schéma* for Paris, to increase accessibility whilst reducing the need for mobility. At the same time, the GLDP was intended to stimulate regeneration and to resist further decline of the central area. The fundamental strategic planning issue for central London is whether there should be decentralization and a weakening of the centre, maintaining or strengthening the centre. Until the late 1960s the increase in services and demand for offices led to net growth, despite some dispersal due to improved roads and, the increase in private transport. Despite the lack of encouragement and in many respects, positive discouragement to job creation and commercial activity in Central London, the demand for development was so great that the centre continued to expand. Office floorspace in Central London expanded from 13,623,200 sq. m in 1957 to 16,684,700 in 1966 to 29% of the UK commercial floorspace. Since then, demand has declined in several sectors, and so too has Central London. By 1974 the proportion of the UK office floorspace in Central London had declined to 25%. The balance between the desire to stimulate the economy of Central London, on the one hand, and to check the intensity of congestion on the other, has varied according to who was in power at the GLC. Labour control from 1981 coincided with high priority for social considerations and the environment. It opposed expansion of the central area south of the river to protect local housing and jobs. However, as planning permission normally lasts for five years, a pool of outstanding permissions, granted at the time of previous political control, to some extent evens out the effects of changes in policy.

Some transport issues in Central London

1 Usage of public transport. Up until 1982 decline in usage was a serious problem. Since then there has been a significant increase in usage and decline in real unit costs and the problem of costs and maintenance of services has been reduced.

2 Peaking of usage. This is a particular problem for rail travel which is largely used for journeys to work.

3 Length of routes. Both underground and buses operate on very long routes, partly as a result of the size of London. For the buses in particular, this exacerbates the problems of timekeeping.

4 Staff shortages. The buses have suffered from staff shortages particularly in the inner areas and the centre. In the first half of the 1980s the GLC decentralized the locations from which operations took place because of the better staffing situation in outer London.

5 Priority for public transport. Priority for buses of course means some delay to other traffic and the dilemma is where to strike the balance of priorities. So far they have been quite limited in scale in London. By 1985 there were 220 bus lanes in London extending over 61 km or about 2% of the bus network. However, as Buchanan and Coombe (1973) pointed out, bus lanes can be quite effective even if small in scale. They calculated that bus lanes on 2% to 3% of the network could reduce delays to buses by 30% whilst if comprising 20% of the network delays would be reduced by 60%.

6 How much car parking should be allowed/provided? All on-street car parking is now controlled in Central London (Table 9.7).

7 Distribution *within* the city centre becomes a problem in cities as large as London or Paris compared with other cities in this book. In

Table 9.7 Central London parking.

	Number of spaces
On-street meter	14,389
Resident	11,690
Other	5,584
TOTAL	31,662
Off-street local authority	11,090
Private	20,500
TOTAL	31,590
Private non-residential	60,000
TOTAL	123,253

Source: Greater London Council (1985).

Fig. 9.8 Oxford Street, like many other parts of Central London, relies heavily on public transport.

Fig. 9.9 Blackfriars Railway Station, London, an important terminal for commuters to the City. The pub survives, in a rather incongruous setting.

London, distribution is largely carried out by the underground and the buses. Peripheral railway terminals are a problem. Few British rail routes cross the city centre and so interchange with the bus or underground is necessary. There have been many proposals for cross-city railways, several of them using the presently disused Snow Hill Tunnel linking the Southern Region services from Blackfriars to Midland Region services at St Pancras (Greater London Council 1979). Other schemes have been proposed to connect the Western Region services at Paddington with the Eastern Region at Liverpool Street and Southern Region at Victoria and London Bridge using tunnels under the central area.

8 Lack of a hierarchy of roads and lack of potential for environmental management. The Central London area traffic control scheme was introduced into the Fulham and Kensington area in 1967 and extended to the West End and the city in 1973. Road capacity was increased by 6% (Wardrop 1975). The scheme also allows traffic controllers to clear roads for emergency vehicles and to direct traffic away from accidents and to give priority to buses and TV surveillance as required. It was calculated that accidents in West London were reduced by 14%.

Fig. 9.10 The Barbican, city of London. Large scale housing close to the centre is not common in British cities but this is an exception.

Central Paris

Central Paris is often rather more widely defined than is Central London. The *Ville de Paris* within the *Boulevard Périphérique,* had a population of 2,176,000 within 105 sq. km in 1982. This compares with a population of 2,291,000 in the 1975 census, but is still a density of 20,724 persons per sq. km compared with 7,790 in Central London. The *Ville de Paris* is a separate *département* with its own *SDAU,* although it does not have a *préfet.* Instead, the Mayor of Paris is an important political figure, as was demonstrated in March 1986 when the Mayor, M. Chirac, became Prime Minister.

Since the 1950s, renovation and redevelopment of the *Cité de Paris* has been accompanied by greater overall building density. For every 100 sq. m demolished, 267 sq. m has been built with an overall increase of 12.3% in floor space.

Although the population has fallen from 2.85 million in 1954, the number of jobs in Central Paris is falling only slightly and is now around 2 million, almost half of which commute in. There has, however, been a change from industrial jobs to service jobs with consequences for commuting. Thus permission was granted to build 1.7 million sq. m of office space between 1969 and 1974 (Bateman and Burtenshaw 1983), mostly in the West End on both sides of the Champs Elysées, Montparnasse and the Gare de Lyon. Just outside Central Paris thus defined to the north west is La Défense, completed in the mid 1980s with 1.5 million sq. m of office floorspace.

Like London, there is a ring of main line railway terminals around the central core – Gare du Nord, de l'Est in the north-east, St Lazare in the north-west, Montparnasse in the south-west, Lyon and Austerlitz in the south-east. These enclose the part of Paris built up before 1840 – the ten inner *arrondissements.* Much of the *métro* network is still confined within them. Also like London, they pose problems of connection, indeed perhaps worse than in London. The lack of an equivalent of the London Underground Circle Line on the Paris *métro* and the lack of direct connection between the Gare du Nord and Gare de Lyon are problems.

Specialization within this area is very significant in determining the pattern of journeys to work. Thus finance is concentrated in the first, eighth and ninth *arrondissements,* the university in the fifth and sixth and high quality shopping in the first and eighth for example. Further out in eastern Paris, the eleventh, twelfth, nineteenth and twentieth *arrondissements* are in mixed industrial/-residential uses, for a long time some of the poorest and most environmentally deprived districts of Paris.

The 35 km of the *Boulevard Périphérique* surrounding the *Ville de Paris* was completed in 1973. Being much closer to the city centre, it

Fig. 9.11 Châtelet les Halles, Paris, the centrepiece of Parisian local transport underneath this vast redevelopment area of the 1970s and 1980s.

has a very different function to the M25 around London. Inside, the rings of interior and exterior *boulevards* do not act as distributors As Hall (1977) points out, the famous *boulevards* created by Baron Haussmann, *préfet* of the *Département du Seine* from 1853 to 1870, are not the boon to traffic planning that they might appear to be. There is very little off-street parking and the extensive commercial frontages create much demand for kerbside stopping. Even more serious, many *boulevards* converge on *ronds-points* or bottlenecks as they might be uncharitably translated. However, because many *boulevards* run parallel to the pre-existing streets, they have allowed an extensive one-way traffic system. Unfortunately, the *boulevards* connect badly to the main line railway stations and so too does the *métro*. An urban *autoroute* was completed along the right (northern) bank of the Seine connecting with the *Boulevard Périphérique* in 1967, but the proposals for the left bank autoroute have since been abandoned, partially built. An excellent account of road and planning proposals is contained in Evenson (1979).

Within Britain and France, London and Paris are unique in terms of their public transport infrastructure, in kind as well as in scale. In such large cities, suburban railways and underground railways can draw on a large population to improve their viability (although the level of subsidy in Paris might suggest otherwise) and in turn, the heavy demand for local transport makes such high-capacity networks close to being a necessity. The large German cities such as Frankfurt and Hamburg, closer to British and French second-tier cities in size, have transport networks more closely resembling Paris or London in kind if not in scale. It is by no means sure that their smaller size is the cause of their higher levels of subsidy in comparison with London. More likely, the relatively low level of subsidy in London is a characteristic shared with other British cities rather than one allowed by its large size.

In both cities, politicians and planners have retreated from large-scale urban road building. Seeing the right bank expressway in Paris, some would say not far enough. The advantages of size and being national capitals have been such that there has been no need to be particularly accommodating towards the car for economic reasons.

10 Redditch

Redditch was designated a new town in April 1964, the original target population of 90,000 being an increase of 58,000 on the existing population. The Master Plan (Wilson and Womersley 1966) was the first in the UK to design for overall mobility needs rather than becoming preoccupied with only one or two modes (Potter 1976), despite the target population being well within the limits of unrestricted movement by the private car (very large centres could not function without high-density mass transport).

The design was to provide for the possibility of a change in emphasis from private towards public transport. Pedestrian routes have been separated from fast-moving vehicles. The designs have allowed pedestrian access as an important mode of transport: the extent of housing areas has been determined by walking distance from local services. Together with recognition of the need for public transport, this reflected another important design principle of the Master Plan – flexibility. In this case it reflected the possibility of insufficient funds for unrestricted road-building and the need for choice in transport.

The new town design consists of a series of linked district centres on a public transport spine connecting them to the town centre. Density varies from 165 persons per hectare (75 persons per acre) near the bus stops to 62 persons per hectare (25 per acre) at the periphery of the town with an average of 124 per hectare (50 per acre). Employment land uses have been distributed to disperse peak-hour traffic demand. 'Residential units' have been designed based on pedestrian access to services. These facilities have been grouped along bus-route corridors rather like the way in which they group along radial roads in old cities, but with the nuisance of heavy traffic separated out. Main roads with heaviest vehicle flows are at the edges of these units, thereby improving the environment and safety within. There are 10 km (6 miles) of routes designated solely for buses and a further 24 km (15 miles) of non-reserved busways. Services on the first main public transport loop, introducing the high quality service envisaged in the Master Plan (Reddibus), were started in March 1976.

By 1981 Redditch was essentially complete, with a population of about 70,000. There was a service frequency of 10 minutes, since reduced to 5 minutes. Little subsidy has been needed. By contrast, Milton Keynes, a low-density new town (overall 22 persons per hectare or 9 per acre) with dispersed traffic-generating activities, in 1980 needed a subsidy of £735,000 for a 30-minute frequency on the 15 bus routes.

However, a household survey carried out by Redditch Development Corporation in 1978 showed that the usage of buses by residents in areas served by the busways was not significantly higher than in other parts of the town. Also, residents perceived the reliability of the busways as being no greater. Bus speeds, averaging 25 km per hour (15.6 mph) are, however, quite high for an average distance between stops of about 1 km (0.62 mile). The main services on busways run an equal frequency at morning and evening rush hours as they do at the rest of the day, perhaps implying a lack of use for work journeys. In the survey of 1978, 11% of journeys to work were by bus.

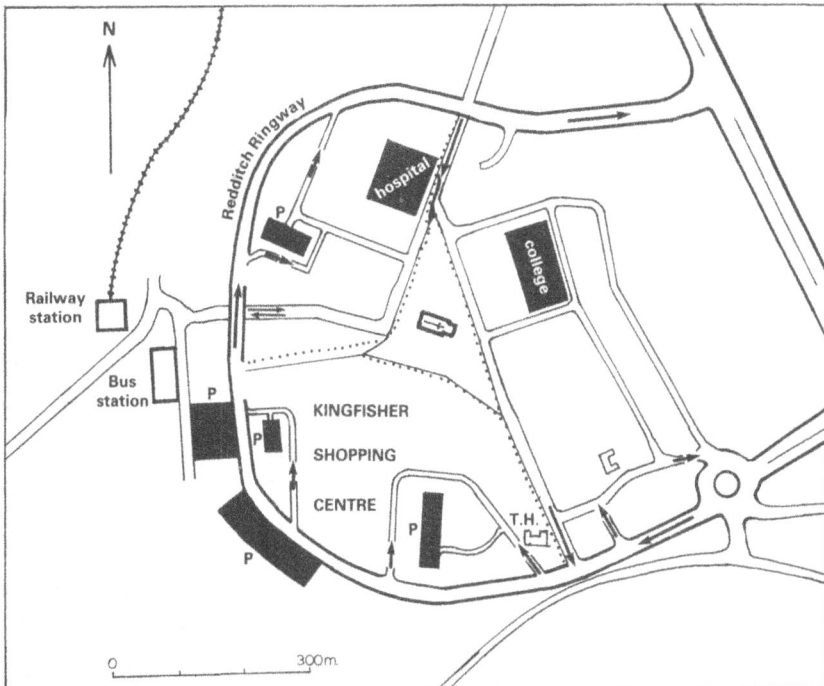

Fig. 10.1 Redditch town centre.

Fig. 10.2 Busways connect to the district centre at Redditch.

Fig. 10.3 Buses and other equipment used are fairly standard.

Fig. 10.4 The central bus station at Redditch connects to the Kingfisher shopping centre.

The Redditch Ringway, connected directly to the town's primary road network, encloses the town centre. Within it there is a distinction in uses between the shopping area to the south-west, where the Kingfisher shopping centre occupies one quarter of the whole town centre, and the civic and cultural facilities to the north and east. Using one-way restrictions, bus-only lanes and pedestrianization, the town centre has been divided into three cells, each accessible from the Ringway, but not directly accessible to each other by the private car. The three cells are: 1) the Kingfisher shopping centre, town hall, library and market in the south 2) Smallwood hospital to the north; and 3) Redditch College, police station and the law courts to the east. In the centre, separating the cells are a pedestrian area and St Stephen's Church. The railway station and bus station are just outside the Ringway to the west. The bus station is connected to the Kingfisher shopping centre by escalator (Fig. 10.4).

Before designation as a new town in 1964, Redditch was an established town of 32,000 population. Since then it has been expanded southwards and eastwards across the River Arrow, structured around a new primary road network. The opportunity was taken to restructure the town centre in its existing location using principles of ringway and traffic cells. Contrary to the majority of new towns of the time, it was envisaged that public as well as private transport would continue to be both necessary and desirable. The other new town of the time, embracing similar sentiments, was Runcorn (Ling 1967).

Whereas Redditch has been expanding in population with opportunity to restructure the town centre at the same time, in the next chapter we shall turn to planning in Burnley, a town presently of similar size, but which has been declining for several decades.

Fig. 10.5 The low density of housing at Milton Keynes has been largely responsible for the much poorer and more heavily subsidized bus services than in Redditch and Runcorn.

11 Burnley

Forty-five kilometres north of Manchester, Burnley and neighbouring small towns of Padiham and Hapton form a local authority which at the time of the 1981 Census had a total population of 94,078 (1971: 96,592) within 117.6 sq. km. Public transport is provided by the Burnley and Pendle Joint Transport Committee covering Burnley Borough and the neighbouring district council of Pendle. Pendle District comprises the two main towns of Nelson and Colne together with neighbouring small towns of Brierfield, Barnoldswick, Trawden and Earby which together had a population of 85,744 in 1981 (1971: 85,416) within an area of 168.2 sq. km.

Burnley grew up during the nineteenth and early twentieth centuries as an industrial town based on cotton weaving. For most of the twentieth century, weaving has been in decline and has now almost disappeared, being partially replaced by a variety of light and general engineering firms, many in former weaving premises. A steady decline in population has accompanied that of the weaving industry. Particularly from the late 1950s until the late 1970s, much high density terraced housing, mostly within 2 km of the town centre, was replaced by lower density housing on the same sites and on the edge of town. Thus the declining population, changes in physical form of the town, increasing unemployment in line with national trends and, above all, the increase in car ownership, all coincided to make public transport more difficult to provide at an economic rate.

Burnley and Pendle are still districts with a relatively high proportion of lower paid jobs (Table 11.1).

Car ownership levels are relatively low. A total of 49.9% of households in Burnley have no car and 45.5% in Pendle compared with 39.5% in Britain as a whole. The use of the car for journeys to work suggests that those who have one use it for this purpose (Table 11.2).

Table 11.1 Distribution of workers in Great Britain.

	Employers, managers, professional workers (SEGs 1–4) %	Semi-skilled and unskilled workers (SEGs 10 and 11) %
Burnley DC	10.39	25.97
Pendle DC	10.80	25.00
Great Britain	14.25	18.46

Source: UK Census 1981, County Report for Lancashire, National Report, Part 2, Table 46.

Table 11.2 Mode of travel to work in Burnley and Pendle (%).

	Car	Bus	Train	Foot	Other
Burnley DC	51.3	18.1	0.6	22.3	7.8
Pendle DC	51.5	10.7	0.5	29.6	7.7
Birmingham	43.9	33.8	3.0	12.6	6.7
Great Britain	50.2	16.0	5.7	15.7	12.4

Source: UK Census 1981, County Report for Lancashire, National Report, Part 2, Table 44.

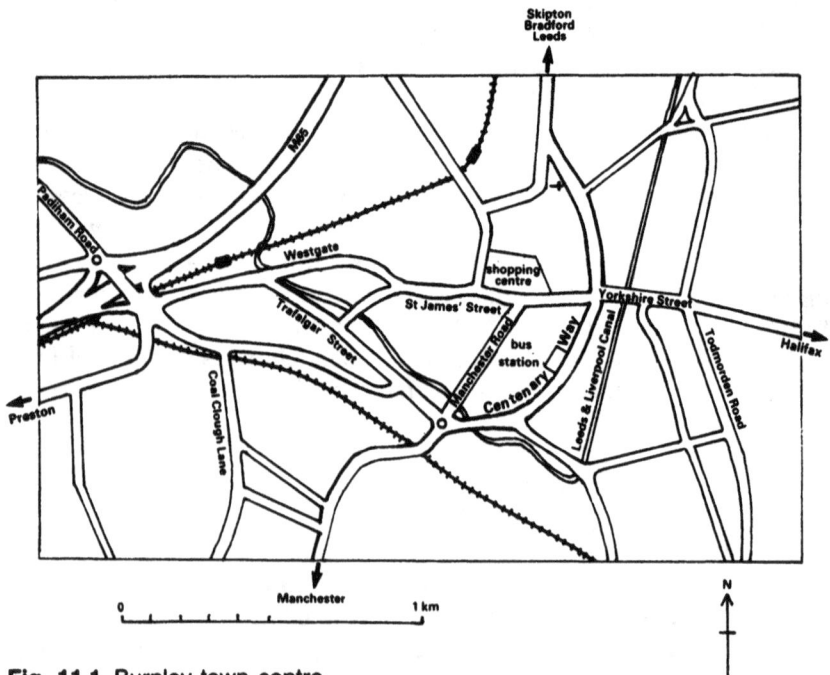

Fig. 11.1 Burnley town centre.

Fig. 11.2 Burnley. Much of the town centre is a product of the 1960s. Access for the car was made much easier, but the town also built a bus station superior to those in many large cities.

Apart from taxis, local public transport is provided almost exclusively by buses. The Burnley and Pendle Joint Transport Committee operates 67 single and 29 double-decker buses. There are also services to neighbouring towns by other local authority transport undertakings – Halifax, Rossendale and Hyndburn (Accrington) – and longer-distance services operated by the National Bus Company and its local subsidiary, Ribble.

Despite the adverse conditions and little restraint on the private car by way of parking restrictions or traffic congestion, receipts from passengers covered 82.13% of turnover in the year ended 31st March 1984 (Burnley and Pendle Joint Transport Committee 1984). Fares were only moderately high. For example, at 31st March 1984, a journey of 4.8 km cost 45p, 30% to 40% above those in the West Midlands for example. Services were, however, generally poorer. Thus in Burnley and Pendle, there is one bus to 1,873 people whilst in the West Midlands there is one to 1,204. In some of the outlying villages, which are not remote compared with more rural parts of the country, services are sparse. The village of Barley for example, is served by buses on three days per week to Clitheroe 8km to the west, Burnley 8 km to the south and Nelson 5 km to the south-east.

In these towns of north-east Lancashire, public transport has little influence in leading development of the urban area. It is above all a

Fig. 11.3 Many other services as well as those of Burnley and Pendle Joint Transport use the bus station.

social service provided so that those who do not have private transport can gain access to facilities whose location has been influenced by decentralization policies favouring greenfield sites and by the increase in availability of the private car.

Decentralization has been perhaps most conspicuous in the form of peripheral housing development since the mid 1950s. Other significant examples are the replacement of inner area jobs by peripheral industrial estates, particularly in the north-east of the town and some factories in the south, and decentralisation of schools. Thus three large secondary schools were built in the 1950s and early 1960s on the western boundary of the former Burnley County Borough and others were built on the eastern and south-eastern fringes during the same period. At the same time there has been a concentration of shopping in the town centre, particularly supermarkets, replacing local shops. Probably initiated by changes in marketing methods, this trend has been helped by the increasing availability of the car and has perhaps contributed to the maintenance of bus services.

Unlike Redditch, there has been no overall restructuring of the town-centre, rather piecemeal improvements. The town centre by-pass (Fig. 11.1) was opened in the early 1960s and takes much of the through traffic from the north and east (Yorkshire) and south (Manchester) away from the main shopping area along St James' Street. No major streets have been pedestrianized although there are substantial pedestrian areas around the Thompson Sports Centre/

Fig. 11.4 With plenty of town centre parking and many customers within walking distance, Burnley centre is probably not as dependent on public transport as large cities. The buses are more significant as a service to customers than as a support to the centre.

Fig. 11.5 The few dozen houses and farms which make up the village of Barley near Nelson are connected to local towns by bus on three days per week. Being in very pleasant upland countryside, there are extra services for visitors in summer.

Library immediately south of the bus station (Fig. 11.1) and the shopping precinct off St James' Street (Fig. 11.3).

Conclusions

Generally, the inter-relationships between public transport and urban planning become stronger with increasing city size. Although the inter-relationships between land use and public transport are complex and not readily measurable, as a general rule local railways have more influence on land-use patterns than do buses. Certainly in towns too small to support a local railway, the transport effects are different from those in larger settlements. In a town such as Burnley, public transport appears to have had much less effect on decisions in land use planning than in towns with local railway networks. The buses have, however, been important in permitting, or at least giving the impression to planners and politicians, that there is freedom in choice of land for development from a public transport point of view. It is important not to confuse freedom of choice of site, permitted by using buses, with lack of influence on development decisions. In taking decisions on the location and density of development, buses have been perceived as being able to accommodate development whose form is compatible with other criteria. Their influence on building has been seen to be permissive.

The higher densities and more intense land uses associated with larger cities gives more incentive for pedestrian and other environmental improvements. Conversely, the reverse situation in smaller settlements makes environmental improvements more easily implemented. On balance, most environmental improvements associated with public transport do not appear to be closely related to settlement size or centrality. They are mostly influenced by local architectural characteristics and urban design characteristics such as street layouts, which are quite independent of settlement size. Traffic measures such as bus-only lanes are an exception.

So the effects of increasing size, centrality and density on public transport and urban development are that the economic functions of public transport in supporting activities become more significant and that rail transport is more likely to be chosen with its associated effects on urban development. The environmental role of public transport perhaps increases too. There are many small towns where the environment is under threat, due to heavy through traffic for example. It is in larger settlements where the threat can be addressed by substituting public for private transport.

12 Could public transport revive the city centre?

Does the city centre need to be revived?

Loss of shopping, jobs, poor environment, crime against property and person are some of the problems associated with the city centre and which have received much attention in recent years. They are all very clearly expressed in the project reports for the new French *métros* and by the consultants and the local transport authorities in the feasibility studies for the light rapid transit proposals in Britain. In France, several thousand million francs have been invested by central and local government and local taxpayers, largely justified by the assumptions that these problems exist and that *métros* will help alleviate them. In Britain, smaller sums of money have been paid to transport consultants to prepare reports to explain the problems and the potential of light rapid transit to help solve them. Birmingham, Manchester and Sheffield in particular have all taken the potential of light rapid transit very seriously (for a recent review of proposals, see Reed 1986).

Much of the shopping lost from centres has been replaced or simply transferred to cheaper premises a short distance away in older parts of the city. For food and other bulky goods where those who have a car are likely to wish to use it, some shopping has transferred to out-of-town centres where parking is easier. This kind of shopping has perhaps been more competitive with the suburban high street than the city centre but the net result has been similar – decline in established centres has coincided with expansion in locations more suited to changing requirements, particularly for access by car. Decentralization to motorway and main road orientated locations peripheral to built up areas has no doubt contributed to a reduction of choice for those who do not have a car and has perhaps made car ownership less voluntary to some of those who do. Decentralization from city centre to inner city or suburbs has been more of a problem to cities' finances than to individuals. Those municipalities who have lost rates revenue have made more noise than those who have gained. Redrawing boundaries seems easier and cheaper than building light rapid transit.

Loss of shopping, commercial activity and jobs from the city centre has been at least partly, and for some sectors largely, a transfer rather than an absolute loss. Improved public transport would probably help to transfer some of it back again and to transfer substitute activities from the inner city to the centre. We may question whether some of these transfers are rather immaterial, whether their occurrence under market forces should be arrested and whether any investment in public transport should be made to help them along.

The city centre is the location of some facilities which could thrive nowhere else. If a city can maintain only one theatre, concert hall, art gallery or only one or two department stores, it makes sense for everyone that they should be close together. Transport can serve one location more efficiently than several, even if they are strung out along a single route. Visits can be multi-purpose. This is the fundamental reason why cities need centres but it does not mean that change or even decline of some activities should be resisted. Cities need centres to maintain and develop the unique facilities that can only be there, and in developing these facilities other non-unique activities will be attracted as well. However, there has not always been a clear definition of which activities are essential and which could be decentralized. Basically, planning exists to restrain the *undesirable* effects of the market, not *all* of them.

Can public transport save the city centre?

The problems caused by loss or, perhaps more accurately, the transfer of shopping, jobs and commercial activities may not be as great as has sometimes been implied in feasibility studies for light rapid transit. Nevertheless they are real in many cities and potential in even more. Environmental deterioration and increased lawlessness are even more clear.

Shopping, entertainment, employers and other city-centre users already rely on public transport to bring in up to a half or two-thirds of those travelling to them. Public transport already is essential to maintain the form of practically all large cities in western Europe. What many city authorities would like to know is whether further investment in public transport would result in it being more effective or whether all of those likely to use public transport are already doing so. Buses are the cheapest way of fulfilling social aims. Should more expensive local railways be built on the grounds that they will go further to meeting economic as well as social aims?

Some travellers on the new French *métros* in Marseille, Lyon and Lille have been attracted from their cars. They have each made some contribution, so far rather limited, to maintaining the shopping and commercial activities of the city centre. In each case, the *métro* has been developed as part of a restructuring of public transport, involving changes to fare structures, priority measures for buses, coordination of timetables, pedestrianization and other measures to improve the environment. The *métros* have certainly been an important part of the package, but it cannot be known whether the remainder would have been effective without them.

In West Germany, local railways have been more numerous and more extensive, and were developed five to ten years earlier than in France. Land use/transport planning has been taken more seriously and applied more rigidly. There has been no equivalent of the Hamburg Density, Model for example, in France or Britain. No doubt War-time destruction has contributed to the willingness and ability to plan the whole of the city and city region as an entity, and to put into practice plans involving basic restructuring. The partition of Germany in 1949 has also influenced city centre development. Some cities near the boundary with East Germany lost much of their hinterlands. The annexation of Berlin allowed second-tier cities to develop capital city functions. The *Mineralölsteuer* from 1967 onwards has been used to finance *U-Bahn* and *S-Bahn* building, but it has not been the only factor helping the city centre. Bremen is one of the few large West German cities without a *U-Bahn* or *S-Bahn* in the centre, and this has not put it to any perceptible disadvantage. Perhaps the trams have been sufficient.

The results of recent studies on light rapid transit in Canadian cities are not incompatible with those in France and West Germany. Lines have recently been opened in Edmonton (opened 1977: 10 km, 8 stations) and Calgary (1981, 12.4 km, 11 stations). The impact on residential and commercial development and densities has so far been modest, despite incentives from the city authorities (Cervero 1984). Both Calgary and Edmonton have introduced zoning bonuses and other incentives to aid development around the stations. Up to 80% increase in floorspace is allowed within 400 metres of most of the stations. In Calgary, parking requirements for new development were reduced by up to 80% and money payments were allowed in lieu. Downtown parking has declined slightly to what by European standards would be still a very high level of 33,000 places, despite an increase in office floorspace. Costs per passenger have risen steadily since their opening, although in Calgary public transport journeys to the city centre have risen from 34% in the early 1970s to 45% in 1985. However, developers seem to have had some confidence in the new light rapid transit. In Edmonton they paid the City $1,700 per gross developable acre and donated 12 acres of land for the light rapid transit right of way. As in the case of San Francisco's Bay Area Rapid Transit network, most development in Calgary seems to have been in anticipation of the new railways rather than following them. Gomez-Ibanez (1985) concluded that light rapid transit had increased capital and operating costs in exchange for only a small gain in ridership, and that improvement in buses may have been a more cost-effective way of achieving the same results. On the other hand, the downturn of the oil-based economy of Alberta has thwarted the success of the light rapid transit. Railways have been shown to have a greater effect on land uses when the local economy is expanding, as was the case in Toronto.

In Newcastle, Howard and Davies (1986) concluded that there was no evidence that the metro had increased the total trade of the City Centre, although there had been some redistribution of commercial activity. The Structure Plan Annual Report (Tyne and Wear County Council 1985) noted little change in retail shopping in Gateshead, North Shields, South Shields, Jarrow, Wallsend or Whitley Bay since the opening of the metro.

So it appears that light rapid transit alone will not save a dying city centre. Together with other public transport and environmental measures it might revive a centre which is wilting and will certainly support and improve the environment of one that is thriving. On purely economic criteria the balance of evidence will usually be against building local railways except where the local economy is expanding.

A serious doubt about light rapid transit in most contemporary West European cities is that they are flying in the face of market evidence. For more than a decade, car parking has been one of the principal attractions of the decentralized locations to where city-centre activities have migrated. Out-of-town shopping centres, DIY stores and agents for office accommodation have all promoted parking as one of their main selling points, and yet city councils are talking about offering not parking, but light railways. Certainly some of these activities are not competitive to the city centre. Car parking is not so essential to those browsing around a boutique as to those intent on loading up with a hundredweight of groceries. Nevertheless, there is an understandable desire from those with a car to use it. The railways will have to be attractive to prise motorists away from their cars, or the city centre will have to be so attractive that they will put up with using what to them is not the best method of transport at their disposal.

One of the main arguments to support light rapid transit is that, although the financial benefits of light rapid transit have sometimes been over-extolled, the costs too have been exaggerated. The accounts that we see in project reports and elsewhere are the transactions which take place on paper, and to a considerable extent, nowhere else. They include employees' income tax and other taxes, National Insurance contributions of both employer and employee and many other 'costs' which are not a reflection of resources committed to the project. As the main 'cost' of most development projects is labour, the figures we see give a particularly inflated impression of resource commitment in times of high unemployment.

Amongst the clearest benefits of light rapid transit are environmental ones. Light railways can bring in people with less environmental damage than buses. If they persuade people away from their cars, so much the better. Usually the improvement to the environment would be considerable in only a few streets where buses are a significant proportion of traffic such as New Street or Corporation Street in Birmingham, but in many cases these are amongst the most important in the city in terms of the perceptions of visitors and residents. Although such benefits are very likely to be reflected in financial terms in city-centre activities, their effects are not easily demonstrable and are liable to be overlooked. Also usually overlooked are the effects on those who continue to travel to the city centre. Presumably those who would put up with the city centre even before any improvements brought by improved public transport will feel the benefits. Their consumer surplus from travelling to the centre will have risen.

So there are at least two good arguments for investing more in public transport to serve the city centre – more than what is necessary to achieve the social objectives of public transport of providing access for those who do not have private transport. Firstly, the costs have been grossly exaggerated and ambiguous; secondly, there are environmental benefits which have not always been fully appreciated and which with the increase in the service sector, tourism, national and international conferences, are becoming more highly valued.

The city centre environment

In cities relying on buses for public transport, investment in light rapid transit would invariably allow environmental improvements, but generally in quite a limited way beyond that already achievable with buses. Environmental improvements per pound sterling invested in light rapid transit must be very low compared with that achieved with buses. Rail-based public transport certainly helps the implementation and operation of pedestrianized streets but is not essential for them to be successful. There are many examples in Britain and France where buses operate up to the edges of successful pedestrian zones. Where they are allowed within otherwise pedestrianized areas such as Bull Street or High Street in Birmingham they are more intrusive than where trams only are allowed, as in Bremen.

However, all this does not apply to the same extent to buses in special lanes or on busways. Busways are generally a cheaper alternative to the *métro* and may be just as effective for all but the highest levels of demand. They are also more easily converted back to other road uses if expectations of demand are not fulfilled and are more easily integrated into the suburban bus network. Busways could be planned to involve fewer changes of transport than is usual with suburban railways where many passengers have to be brought to them by bus.

Like local rail transport, pedestrianization has been carried out on a large scale in West Germany. Monheim (1974) estimates that by 1971 there were already 134 town-centre pedestrianization schemes and by the end of 1973 this had increased to 220. This is estimated to have further increased to 340 by the end of 1976 (Frankfurt am Main, Dezernat Planung, 1977). Also, even in city centres, pedestrianization is not confined to shopping streets. Areas around town halls, libraries and other civic uses are in many cities given over to pedestrians. In Germany, traffic restriction has been seen as a way of improving the attractiveness of the city centre for housing (Frankfurt am Main, Dezernat Planung, 1977).

So local railways or trams offer some advantages over buses in the improvement of the city-centre environment. They have a higher capacity and bring in more people for less noise and other intrusion. Buses themselves generally cause a nuisance in only a few streets, but usually important ones, and this is where rail transport offers the advantage. Rail transport is also usually more acceptable in streets which are otherwise public transport only. However, it is quite reasonable to question whether these benefits justify the extra costs of rail transport. One characteristic shared by all the cities with a recently-built rail network (except Tyne and Wear) is that public transport subsidies are high. The proportion of public transport costs met from fares is generally less than half that in cities depending on buses.

City centre planning policies have reflected a desire to progress towards four main objectives:

1 to give accessibility;

2 to maintain and enhance the environment;
3 to cause a minimum of disturbance to existing users;
4 to stimulate the economy of the centre.

Environmental improvement and noise reduction tend to be incompatible with easy access, whilst economic objectives may be compatible with both good access and improvement of the environment. City-centre planning and public transport policies reflect the outcome of these four groups of considerations. Many cities have tried to reconcile the incompatibility of the environment and good accessibility by one or both of two main principles. Firstly, they have allowed designated areas to be environmental 'write-offs' for the sake of accessibility, whilst allowing the environment to take precedence elsewhere. Ring roads and their environs are the main way in which this has been put into practice. Secondly, they have spent more than the minimum needed to provide a public transport service for social purposes in building a local rail network, more expensive than providing a comparable service with buses but with some further environmental benefits. West German cities have gone a long way to apply both of these principles. Few British cities have gone as far with the environmental 'write-off' solution as have Birmingham and Coventry and few French cities have got far with either, except those with new *métros*. Disturbance to existing users has taken place to a much greater extent in West German cities which together with their willingness to spend far more on public transport has meant that more planning has been implemented. Whether French and British cities will proceed in the same direction is doubtful. The British have chosen more private rather than public spending than have the Germans. Both British and French have valued individual rights so highly that large scale disruption by public works has not been allowed to the same extent and in any case, the Second World War left their cities in a very different state from those in Germany, less ripe for overall basic restructuring.

The quality of life reflected in the environment has become a more valued issue than it was in the boom years of the 1960s. In West Germany, the Green Party has grown steadily to achieve 9% of the votes in the election for the Bundestag in January 1987. More and more members of the public other than those in local interest groups and pressure groups are beginning to appreciate the dangers for quality of life of the city centre as elsewhere, in pursuing purely economic goals. In the 1960s, many city centres were battlefields where private profit often won over public interests. Soon, after some conspicuous mistakes – empty office blocks, large-scale demolition for urban road-building, for example – environmental planning was taken more seriously. Perhaps the next problems building up are social rather than economic.

Appendix

A public opinion survey on the potential for public transport in Birmingham city centre

A questionnaire survey was carried out in November 1985 to provide evidence on the potential for light rapid transit in the city centre and on the closely related field of attitudes towards public and private transport. Thirty-four surveyors at street locations spread evenly throughout the Centre to represent shopping, office and commercial uses interviewed randomly selected respondents.

1 How did you come to the city centre today?

Weekdays 10am to 3.30pm:		Walk	4
Car	54	Motor cycle	3
Bus	112	Bicycle	2
Train	21	Other	3
Walk	7		
Motor cycle	4	Saturdays 9am to 5pm:	
Bicycle	4	Car	33
Other	2	Bus	76
		Train	19
Weekdays 8am to 9am:		Walk	4
Car	34	Motor cycle	0
Bus	70	Bicycle	2
Train	20	Other	2

2 How many times a week do you come into the city centre?

Weekdays 10am to 3.30pm:			Three times	13
Less than once	5		Four times	5
Once	35		Five or more times	93
Twice	23			
Three times	39			
Four times	19		Saturdays 9am to 5pm:	
Five or more times	83		Less than once	23
			Once	22
Weekdays 8am to 9am:			Twice	21
Less than once	4		Three times	11
Once	6		Four times	11
Twice	15		Five or more times	44

3 Where do you live?

Weekdays 10am to 3.30pm:			Outer Birmingham	79
Inner Birmingham	34		Elsewhere	25
Outer Birmingham	107			
Elsewhere	63		Saturdays 9am to 5pm:	
			Inner Birmingham	37
Weekdays 8am to 9am:			Outer Birmingham	77
Inner Birmingham	32		Elsewhere	22

4 Do you own a car?

Weekdays 10am to 3.30pm:			Car in family	36
Owner	87		No	53
Car in family	51			
No	66		Saturdays 9am to 5pm:	
			Owner	64
Weekdays 8am to 9am:			Car in family	21
Owner	47		No	51

5 What was your main reason for coming into town today?

Weekdays 10am to 3.30pm:			Leisure	2
Work	67		Education	10
Shopping	75		Other	15
Leisure	14			
Education	16		Saturdays 9am to 5pm:	
Other	32		Work	10
			Shopping	101
Weekdays 8am to 9am:			Leisure	17
Work	101		Education	2
Shopping	8		Other	6

6a) Do you know of the County Council's proposals for a light rapid transit system?

Weekdays 10am to 3.30pm:		Saturdays 9am to 5pm:	
Yes	137	Yes	79
No	67	No	57

Weekdays 8am to 9am:	
Yes	84
No	52

b) If it was built, would you come into town more often?

Weekdays 10am to 3.30pm:		Saturdays 9am to 5pm:	
Yes	50	Yes	58
No	126	No	60
Do not know	28	Do not know	18

Weekdays 8am to 9am:	
Yes	20
No	107
Do not know	9

c) If no, was it because it would not be near to where you live, or because you do not like rapid transit?

Weekdays 10am to 3.30 pm		Saturdays 9am to 5pm:	
Not near	67	Not near	38
Do not like	18	Do not like	0

Weekdays 8am to 9am:	
Not near	48
Do not like	6

d) How far would you be prepared to travel to it?

Weekdays 10am to 3.30pm:		400 to 800 metres	17
Less than 400 metres	61	More than 800 metres	12
400 to 800 metres	52		
More than 800 metres	8	Saturdays 9am to 5pm:	
		less than 400 metres	28
Weekdays 8am to 9am:		400 to 800 metres	39
Less than 400 metres	50	More than 800 metres	14

7 Do you think there should be more bus-only streets like Bull Street or High Street?

Weekdays 10am to 3.30pm:		No	35
Yes	115	Do not know	27
No	44		
Do not know	45	Saturdays 9am to 5pm:	
		Yes	76
Weekdays 8am to 9am:		No	37
Yes	74	Do not know	23

8 Is there too much or too little car parking in the city centre?

Weekdays 10am to 3.30pm:		Right amount	27
Too much	6	Do not know	39
Too little	123		
Right amount	36		
Do not know 39		Saturdays 9am to 5pm:	
		Too much	12
Weekdays 8am to 9am:		Too little	60
Too much	10	Right amount	24
Too little	75	Do not know	40

9 Is there a need for better transport from one part of the city centre to other parts?

Weekdays 10am to 3.30pm:		Saturdays 9am to 5pm:	
Yes	103	Yes	6
No	56	No	53
Do not know	45	Do not know	22

Weekdays 8am to 9am:	
Yes	50
No	48
Do not know	38

10 How important is it to have a new bus station in the central area?

Weekdays 10am to 3.30pm:		Saturdays 9am to 5pm:	
Very important	55	Very important	31
Fairly important	85	Fairly important	56
Not important	64	Not important	49

Weekdays 8am to 9am:	
Very important	27
Fairly important	62
Not important	47

11 Would you mind telling me which age group you are in?

Weekdays 10am to 3.30pm:		30 to 59	41
Less than 18	8	60 and over	11
18 to 29	86		
30 to 59	87		
60 and over	23	Saturdays 9am to 5pm:	
		Less than 18	9
Weekdays 8am to 9am:		18 to 29	50
Less than 18	9	30 to 59	59
18 to 29	75	60 and over	18

12 Record sex of participants.

Weekdays 10am to 3.30pm:		Saturdays 9am to 5pm:	
Male	113	Male	70
Female	91	Female	66

Weekdays 8am to 9am:	
Male	74
Female	62

	Socio-economic group	Survey (%)	Birmingham (%)
1	Employers in industry, commerce etc – large establishments; managers in central and local government, industry, commerce etc – large establishments	4.0	3.0
2	Employers in industry, commerce etc – small establishments; managers in industry, commerce etc – small establishments	5.5	5.1
3	Professional workers – self employed	7.3	0.5
4	Professional workers – employees	10.7	2.3
5.1	Ancillary workers and artists	2.6	4.6
5.2	Foremen and supervisors – non-manual	1.9	0.7
6	Junior non-manual workers	20.1	7.5
7	Personal service workers	6.9	1.5
8	Foremen and supervisors – manual	0.3	2.9
9	Skilled manual workers	3.9	17.6
10	Semi-skilled manual workers	4.4	13.9

11	Unskilled manual workers	3.7	4.5
12	Own account workers (other than prof.)	4.0	2.8
13	Farmers – employers and managers	0.0	0.0
14	Farmers – own account	0.0	0.0
15	Agricultural workers	0.0	0.0
16	Members of armed forces	0.3	0.0
17	Inadequately described occupations	14.7	30.8

The figures for Birmingham refer to heads of households in the 1981 Census. Those in the Survey are not necessarily heads of households.

Employment categories in the survey compared with the 1981 Census

The survey, of course, can only expect to represent the attitudes of present visitors to the city centre. It may indicate their attitudes towards public and private transport and in particular the light rapid transit proposals, and it may help to estimate any increase or decrease in travel to the city centre by those who already visit it, biased in favour of those who visit it most frequently (as their chances of being interviewed in the survey would be greater). The survey obviously does not include those who might be tempted to start visiting the city centre and is heavily biased against those who do not visit it very frequently. The question therefore arises as to whether those surveyed are typical of Birmingham residents as a whole. If a light rapid transit network was built, or if other changes in the city centre transport tempted more to visit it, the travel habits of Birmingham residents in general may be at least as significant in determining the effectiveness of any new transport measures as the travel habits of previous city centre visitors.

Clearly the sample contacted differs significantly from Birmingham residents as a whole on several important criteria, for example access to a car:

	Car available	Car not available
Survey November 1985	306 (64.3%)	170 (35.7%)
1981 Census (all Birmingham)	50.5%	49.5%

Attitudes towards light rapid transit

At each of the times surveyed, only a minority would travel more frequently if light rapid transit was built. Least effect was expressed by rush-hour travellers, followed by off-peak users, whilst Saturday users, mostly shoppers, expressed greatest willingness to use it. Are these differences great enough to be significant, or are they merely part of a random variation to be expected in any series of surveys?

From the answers to question 6(b):

Observed	Yes	No	Do not know	Total
Weekdays 8am to 9am	20	107	9	136
Weekdays 10am to 3.30pm	50	126	28	204
Total	70	223	37	340

Applying the Chi squared test to see whether the difference bewteen the results for weekdays 8am to 9am and weekdays 10am to 3.30pm is significant:

Expected	Yes	No	Do not know
Weekdays 8am to 9am	$\dfrac{70 \times 136}{340} = 28$	$\dfrac{233 \times 136}{340} = 93.2$	$\dfrac{37 \times 136}{340} = 14.8$
Weekdays 10am to 3.30pm	$\dfrac{70 \times 204}{340} = 42$	$\dfrac{233 \times 204}{340} = 139.8$	$\dfrac{37 \times 204}{340} = 22.2$

$$\text{Chi squared} = \sum \frac{(O-E)^2}{E} = \frac{(20-28)^2}{28} + \frac{(107-93.2)^2}{93.2} + \frac{(9-14.8)^2}{14.8}$$

$$+ \frac{(50-42)^2}{42} + \frac{(126-139.8)^2}{139.8} + \frac{(28-22.2)^2}{22.2}$$

$$= 2.2857 + 2.0433 + 2.2730 + 1.5238 + 1.3622 + 1.5153$$

$$= 11.0033$$

With two degrees of freedom, the chance of there being no

significant difference between the answers on-peak and off-peak is less than 0.005.

Similarly, the Chi squared test can be used to test whether the difference between the results for weekdays 10am to 3.30pm and Saturdays 9am to 5pm is large enough to be significant:

Observed	Yes	No	Do not know	Total
Weekdays 10am to 3.30pm	50	126	28	204
Saturdays 9am to 5pm	58	60	18	136
Total	108	186	46	340

Expected	Yes	No	Do not know
Weekdays 10am to 3.30pm	$\dfrac{108 \times 204}{340} = 64.8$	$\dfrac{186 \times 204}{340} = 111.6$	$\dfrac{46 \times 204}{340} = 27.6$
Weekdays 9am to 5pm	$\dfrac{108 \times 136}{340} = 43.2$	$\dfrac{186 \times 136}{340} = 74.4$	$\dfrac{46 \times 136}{340} = 18.4$

$$\text{Chi squared} = \sum \frac{(O-E)^2}{E} = \frac{(50-64.8)^2}{64.8} + \frac{(126-111.6)^2}{111.6} + \frac{(28-27.6)^2}{27.6}$$

$$+ \frac{(58-43.2)^2}{43.2} + \frac{(60-74.4)^2}{74.4} + \frac{(18-18.4)^2}{18.4}$$

$$= 3.3802 + 1.8581 + 0.0060 + 5.0704 + 2.7871 + 0.0087$$

$$= 13.1105$$

With two degrees of freedom the probability of there being no significant difference between the answers for the weekday off-peak period and Saturdays is less than 0.005. Therefore we can be more than 99.5% sure that those interviewed on Saturdays were more likely to travel more to the city centre using light rapid transit than those interviewed at the weekday off-peak period. We are also more than 99.5% sure that those interviewed at weekday off-peak periods would be more likely to increase the number of journeys to the city centre than those interviewed weekday on-peak.

The answers to Question 5 show that there is a very clear relationship between the proportions of journeys to work and shopping according to the time of the survey and day. Are those travelling to the city centre for shopping more likely to increase the number of visits than those travelling to work, if light rapid transit was built?

Analysis of the original questionnaires (not the results presented above) revealed the following:

Question 5

		Shopping	Work	Total
Question 6(b)	Yes	54	23	77
	No	108	141	249
	Total	162	164	326

For example, 54 shoppers said 'yes' to question 6(b), 108 said 'no'.

The Phi coefficient can be used to lest whether the difference between shoppers' and workers' answers is significant.

Dichotomous variable X

		1	0	
Dichotomous	1	a	b	a + b
variable Y	0	c	d	c + d
		a + c	b + d	n

$$\text{Phi coefficient} = \frac{ad-bc}{\sqrt{(a+b)\,(c+d)\,(a+c)\,(b+d)}}$$

$$= \frac{(54 \times 141) - (23 \times 108)}{\sqrt{77 \times 249 \times 162 \times 164}}$$

$$= \frac{7{,}641 - 2{,}484}{\sqrt{509{,}388{,}260}}$$

$$= 0.2273$$

With 324 degrees of freedom, the chance of no significant difference between willingness of shoppers and workers to travel more after light rapid transit is built is between 0.05 and 0.02. Therefore we can be more than 95% sure that shoppers have expressed a more significant willingness to increase the number of their visits to the city centre with light rapid transit than have those travelling for work.

If light rapid transit was built, would those who do not have a car available be more likely to travel than those who do have access to a car?

Question 4

		Not car owner	Owner or car in family	Total
Question 6(b)	Yes	53	75	128
	No	94	199	293
	Total	147	274	421

$$\text{Phi coefficient} = \frac{(53 \times 99) - (75 \times 94)}{\sqrt{128 \times 293 \times 147 \times 274}}$$

$$= \frac{10{,}547 \times 7{,}050}{\sqrt{1{,}510{,}586{,}100}}$$

$$= 0.0900$$

With 419 degrees of freedom, the probability of no significant difference is greater than 0.10, therefore the chance of those without a car being more likely to travel more after light rapid transit than those who do have a car available is less than 90%. No significant difference between car owners and non-car owners has been proved.

Are those who already travel to the city centre by public transport more likely than those who travel by private transport to increase their journeys after light rapid transit is built?

Question 1

		Bus and train	Car, walk, m/c, bicycle	Total
Question 6(b)	Yes	93	34	127
	No	191	100	291
	Total	284	134	418

$$\text{Phi coefficient} = \frac{9{,}300 - 6{,}494}{\sqrt{1{,}406{,}435{,}500}}$$

$$= 0.0748$$

With 416 degrees of freedom, the chance of no significant difference between public and private transport users in their answers to Question 6(b) is greater than 0.10. We are therefore less than 90% sure that public transport users are more likely to increase their journeys to the city centre than are private transport users. No significant difference between public and private transport users has been proved by the survey.

It may be that publicity will increase the likelihood of more journeys to the city centre if light rapid transit is built. The answers to Question 6(a) reveal those respondents who heard about light rapid transit for the first time in the survey and those who previously knew of it. Are those who had previously heard of it more likely to use light rapid transit than the others?

Question 6(a)

		Yes	No	Total
Question 6(b)	Yes	72	48	120
	No	202	99	301
	Total	274	147	421

Phi coefficient $= \dfrac{7,128 - 9,696}{\sqrt{1,454,841,300}}$

$= -0.0673$

This weak negative correlation implies that those who had heard of light rapid transit before the survey were *less* likely to reply that they would travel more to the city if light rapid transit was built. With 419 degrees of freedom, the probability of this correlation not being significant is, however, greater than 0.10. We are therefore less than 90% sure that not knowing of light rapid transit before the survey was correlated with being more likely to travel extra journeys to the city centre if it was built.

Bibliography

Abbott, J. (1985), 'Planning with security', *Modern Railways*, 42, 293–4.

Albers, G. (1977), Städtbauliche Konzepte im 20 Jahrhundert – Ihre Wirkung in Theorie und Praxis ('Town planning models in the 20th century – their effects in theory and practice'), *Berichte Raumforschung Raumplanung*, 1, 14–26.

Albers, G. (1986), 'Changes in German town planning, *Town Planning Review*, 57, 17–34.

Allgemeiner Deutscher Automobil Club (1983), *Dokumentation Parkuhrgebühren/Erhöhung* ('Parking meter charge increases') Munchen.

Bateman, M. and Burtenshaw, D. (1983), 'Commercial pressures in Central Paris', Chapter 11 of Davies, R.L. and Champion A.G. (eds) *The Future of the City Centre*, Academic Press, London.

Baubehörde Hamburg (1985), *Sieh Dir an, wie Hamburg baut, 25 Jahre Informationsfahrten der Baubehörde* ('See how Hamburg has developed: a journey through 25 years of building development'), Freie und Hansestadt Hamburg.

Bennett, R. J. (1983), *The finance of cities in West Germany*, Progress in Planning series, Pergamon Press, Oxford.

Beaufort (1984), *Birmingham City Centre. Attitudes towards pedestrianisation, access and the value of public transport to City Centre retail trade*, Main report, October 1984. Beaufort Research Ltd., Newport, Wales.

Bieber, A. (1985), 'Le rôle des transports en commun dans la planification de l'agglomération Lyonnaise' ('The role of public transport in the planning of Greater Lyon'), *Recherche Transports Sécurité* 5, 5–10.

Birmingham City Council (1980), *Central Area District Plan, topic papers*, Birmingham.

Birmingham City Council (1982), *Central Area District Plan written statement*, Birmingham.

Birmingham City Council (1984), *Central Area Local Plan summary of main proposals*, Birmingham.

Blacksell, M. (1982), 'Reunification and the political geography of the Federal Republic of Germany', *Geography*, 67, 310–19.

Boegner, A. (1983), *Operational demonstration of dual mode buses in the City of Essen*, PTRC Summer School Annual Meeting, Brighton.

Bouffartigue, P. (1985), 'Le métro de Marseille, des premiers projets à l'ouverture de la première ligne' ('The Marseille Métro, from the earliest projects to the opening of the first line'), *Transports, Urbanisme, Planification*, 4, 93–113, Centre d'Etudes des Transports Urbains, Ministère des Transports/Ministère de l'Urbanisme et du Logement.

Breton, J. and Lengacher, J-C. (1985), 'Déplacements, transport et commerce dans le centre de Besançon' ('Journeys, transport and commerce in the centre of Besançon'), *Transports, Urbanisme, Planification*, 5, 37–51, Centre d'Etudes des Transports Urbains, Ministère des Transports/Ministère de l'Urbanisme et du Logement.

Bruton, M.J. (1983), 'Local plans, local planning and development plan schemes in England 1974–1982, *Town Planning Review*, 54, 4–23.

Buchanan, L.M. and Coombe, R.D. (1973), 'Bus priority in Greater London 5: assessment of alternative bus priority strategies for inner London', *Traffic Engineering and control*, 14, 522–5.

Buchanan, M., Bursey, N., Lewis, K. and Mullen, P. (1980), *Transport Planning for Greater London*, Saxon House, Farnborough.

Bundesminister für Raumordnung, Bauwesen und Städtebau (1978), 'Fahrrad im Nahverkehr' ('The bicycle as a means of local transport'), *Schriftenreihe Heft 03,066*, Bonn/Bad Godesberg

Bundesminister für Verkehr (1971), *Konzept zur Verbesserung des öffentlichen Personennahverkehrs* ('Principles for the improvement of local public passenger transport'), Bonn.

Bundesminister für Verkehr (1980), *Gemeindeverkehrsfinanzierungsgesetz* ('Community Transport Finance Law'), Bonn.

Bundesminister für Verkehr (1982), *Verkehr in Zahlen* ('Transport Statistics'), Bonn.

Bundesminister für Verkehr (1984), *Bericht uber die Verwendung der Finanzhilfen des Bundes zur Verbesserung der Verkehrsverhältnisse der Gemeinden für das Jahr 1982* ('Report on the use of Federal financial help for the improvement of local community transport for the year 1982'), Bonn.

Bundesvereinigung der Strassenbau und Verkehrsingenieure (1983), *Strasse und Schiene, Daten und Fakten* ('Road and railway transport, data and information'), Bonn.

Burnley and Pendle Joint Transport Committee (1984), Annual Report, Burnley.

Burtenshaw, D. (1985), 'The future of the European city: a research agenda', *Geographical Journal*, 151, 365–70.

Burtenshaw, D., Bateman, M. and Ashworth G.J. (1981), *The City in West Europe*, John Wiley and Sons, Chichester, New York.

Cahiers Français (1985a), *La population français de A à Z* ('The population of France from A to Z'). 219, January/February.

Cahiers Français (1985b), *Le décentralisation en marche* ('Progress on decentralization'), 220, March/April.

Central Statistical Office (1984), *Annual Abstract of Statistics*, HMSO, London.

Central Statistical Office (1985), *Social Trends*, HMSO, London.

Centre d'Etudes des Transports Urbains (1978a), *Les transports collectifs dans l'aménagement des quartiers nouveaux* ('Public transport in the planning of new suburbs'), Ministère de l'Environnement et du Cadre de Vie/Ministère des Transports.

Centre d'Etudes des Transports Urbains (1978b), *L'organisation des déplacements dans la politiques d'aménagement de huit villes Européennes* ('The management of journeys in the planning policies of eight European cities'), Ministère de l'Environnement et du Cadre de Vie/Minisère des Transports.

Centre d'Etudes des Transports Urbains (1979a), *Les aménagements des axes prioritaires de transports collectifs* ('The design of routes giving priority for public transport'), Ministère de l'Environnement et du Cadre de Vie/ Ministère des Transports.

Centre d'Etudes des Transports Urbains (1979b), *Etudes de suivi des ouvertures des métros de Lyon et Marseille* ('Follow-up studies after the opening of the metros in Lyon and Marseille'), Centre d'Etudes des Transports Urbains, Ministère de l'Environnement et du Cadre de Vie/Ministère des Transports, Bagneux.

Cervero, R. (1984), 'Rail transit and urban development', *J. American Planning Association*, 50, 133–47.

Cervero, R. (1985), 'A tale of two cities: light rapid transit in Canada', *J. Transportation Engineering*, 111, 633–50.

Chambre de Commerce et d'Industrie de Lyon (1983), *Déplacements, moyens de transport lies aux achats* ('Journeys and modes of transport for shopping'), Lyon.

Chapuy, P.M.B. (1984), 'France' in Williams R.H. (Ed), *Planning in Europe. Urban and regional planning in the EEC*, George Allen & Unwin, London.

Clout, H.D. (1982), 'A new France?', *Geography*, 67, 244–50.

Cohen, S. (1981), 'Paris, capital de charme', *Cahiers Français*, 203, 34–7.

Conseil Régional Provence-Alpes-Côte d'Azur (1984), *Votre Region de A à Z*, Marseille.

Conseil Régional Provence-Alpes-Côte d'Azur (1985), *Transports et Communications*, Marseille.

Cowling, T.H. and Steeley, G.C. (1973), *Sub-regional planning studies: an evaluation*, Pergamon, Oxford.

Cullingworth, J.B. (1982), *Town and Country Planning in Britain*, 8th Ed., George Allen & Unwin, London.

Dalmais van Straaten, C. (1985), 'Commerces, déplacements et aires de chalandise dans les centres secondaires: quelques cas de l'agglomération Lyonnaise' ('Trade, journeys and catchment areas of secondary centres: some cases from Greater Marseille), *Transports, Urbanisme, Planification*, 5, 73–84, Centre d'Etudes des Transports Urbains, Ministère des Transports/Ministère de l'Urbanisme et du Logement.

Dalmais, C. and Mazzella, P. (1985), *Métro et Urbanisme. Du suivi à l'anticipation*, Agence d'Urbanisme de l'Agglomération Marseillaise.

Daly, A.J. and Zachary, S. (1977), *The effect of free public transport on journey to work*, TRRL, Department of the Environment/Department of Transport, Report SR338, Crowthorne.

Damesick, P.J., Lichfield, N. and Simmons, M. (1986), 'The M25–A new geography of development?', *Geographical Journal*, 152, 155–75.

D'Arcy, F. and Jobert, B. (1975), 'Urban Planning in France' in Hayward, J.

and Watson M. (Eds) *Planning, politics and public policy: The British, French and Italian experience*, Cambridge University Press, Cambridge.

Dattelzweig, D. (1983), 'Auf der Suche nach Lösungsmöglichkeiten' ('In search of possible solutions'), *Deutsche Verkehrszeitung*, 48, 23rd April.

Davies, R.L. and Champion A.G. (Eds) (1983), *The future for the city centre*, Academic Press, London.

Département des Statistiques des Transports (1982), *Bulletin Mensuel de statistiques* ('Monthly bulletin of statistics'), Ministère des Transports, Service d'analyse économique, Paris.

Department of the Environment (1979), *Memorandum on structure and local plans*, Circular 4/79, HMSO, London.

Department of Transport (1982), *Urban public transport subsidies. An economic assessment of value for money*, HMSO, London.

Department of Transport (1985), *Preparing for deregulation – a guide to arrangements for providing local bus services during 1986*, Circular 3/85 (Scottish Development Department Circular 32/85, Welsh Office Circular 64/85), HMSO, London.

Dersjant, A.W., Thuyl, J.M.V. and Steenbrink, A. (1980), *A study of user preference between trams and buses*, PTRC proceedings, Seminar N, July 1980, 29–36.

Dorsch Consult Ingenieurgesellschaft MbH (1977), *Untersuchung zur Generalverkehrsplanung der Stadt Bremen*, ('Research for the City of Bremen Traffic Master Plan'), Wiesbaden/Hamburg.

Dunn, J.A. (Jr) (1981), *Miles to go. European and American Transport policies*, MIT Press, Cambridge, Massachusetts.

Durand, B. and Pêcheur, P. (1985), 'Evolution des transport urbains', *Transports, Urbanisme, Planification*, 5, 5–21, Centre d'Etudes des Transports Urbains, Ministère des Transports/Ministère de l'Urbanisme et du Logement.

Edwards–May, D. (1979), '*Métro* in Lyon and Marseille, parts 1 and 2', *Modern Railways*, 36, Nos 364 and 365.

Elkins, T.H. (1984), '*S-Bahn* goes west', *Geographical Magazine*, 56, 166–8.

Evenson, N. (1979), *Paris: a century of change 1878–1978*, Yale University Press, New Haven and London.

Eversley, D. (1974), 'Britain and Germany: local government in perspective', in Rose, R. (Ed), *The management of urban change in Britain and Germany*, Sage, London/Beverly Hills.

Farrington, J.H. (1986), 'Deregulation of the British bus system', *Geography*, 71, 258–60.

Fédération Internationale Européene de la Construction de la Communauté Groupe des experts économiques (1983), *Les infrastructures de transport et leur financement* ('Transport infrastructure finance'), Fédération Nationale des Travaux Publics, Paris.

Fédération Nationale des Agences d'Urbanisme (1981), *Urbanisme, Déplacements, Transports* ('Town Planning, Journeys, Transport'), Compte Rendu du Colloque, Lyon, 14 et 15 Octobre, 1981.

Field, B.G. (1983), 'Local plans and local planning in Greater London', *Town Planning Review*, 54, 24–40.

Fournie, A. and Pêcheur, P. (1985), 'Financement du systeme de transport et systeme commercial', *Transports, Urbanisme, Planification*, 5, 141–50, Centre d'Etudes des Transports Urbains, Ministère des Transports/ Ministère de l'Urbanisme et du Logement.

Foulkes, M. (1983), 'Transport for Central London' in Davies, R.L. and Champion, A.G. (Eds), *The future for the city centre*, Academic Press, London.

Frankfurt am Main, Dezernat Planung (1976), *Generalverkehrsplan '76* ('Traffic Master Plan '76'), Der Magistrat der Stadt Frankfurt am Main.

Frankfurt am Main, Dezernat Planung, (1977), *Verkehrsberuhigte Zonen in Frankfurt am Main*, Der Magistrat der Stadt Frankfurt am Main.

Frankfurt am Main, Dezernat Planung (1984), *Generalverkehrsplan '76–'82 Netauswahl, Strasse, Scheine* ('Traffic Master Plan '76–'82, network options, road, rail').

Frankfurter Verkehrs- und Tarifverbund (1976), *FVV in Zahlen* ('Frankfurt Transport Authority in figures').

Frankfurter Verkehrs- und Tarifverbund (1984), *10 Jahre FVV* ('10 years of the Frankfurt Transport Authority').

Freie Hansestadt Bremen (1983), *Flächennutzungsplan* ('Land use proposals plan').

Freie Hansestadt Bremen (1984), *Raum für Fussganger* ('Pedestrian areas').

Freie und Hansestadt Hamburg (1973), *Flächennutzungsplan*, Hamburg.

Girnau, G. (1983), 'Wo kann gespart werden im U- und Stadtbahnbau?', *Der Nahverkehr*, Jan, 8–16.

Girardet, A. (1976), *Der Nahverkehr Probleme und Lösungsatze* ('Local transport problems and possible solutions'), Essen.

Gomez-Ibanez, P. (1985), 'A dark side of light rail? The experience of three new transit systems', *J. American Planning Association*, 51, 337–51.

Greater London Council (1976), *Greater London Development Plan*.

Greater London Council (1979), *London's rail network north of the Thames*, Committee Report LT 136.

Greater London Council (1982), *Transport policies and programme 1982–84*.

Greater London Council (1983), *Travel patterns in London and the effects of recent fares changes*.

Greater London Council (1985), *Transport policies and programme, 1979–84*.

Hajdu, J.G. (1978), 'The German city today: crosscurrents of readjustment and change', *Geography*, 63, 23–30.

Hall, P. (1977), 'The World Cities', 2nd ed, Weidenfeld and Nicholson, London.

Hall, P. (1982), *Urban and Regional Planning*, Penguin, Harmondsworth.

Hall, P. and Hass-Klau, C. (1985), *Can rail save the city?* Gower, Aldershot.

Hall, P., Thomas, R., Gracey, H. and Drewett, J.R. (1973), *The containment of urban England*, vols 1 and 2, P.E.P. (London), George Allen & Unwin (London) and Sage Publications (Beverly Hills).

Hamburger Verkehrsverbund (1981), *The Hamburg Transport Community: duties, organisation, data, perspectives*, Hamburg.

Hamburger Verkehrsverbund (1985), *Bericht '85* ('Report '85'), Hamburg.

Hanappe, O. (1983), 'Financement des transports urbains et processus de planification', in CETUR, Ministère des Transports (Eds), *Transports, Urbanisme, Planification*, Bagneux.

Hansestadt Lübeck, der Senat der (1973), *Zieldiskussion und alternative Modelle zur Sanierung der Lübecker Innenstadt* ('Objective evaluation and alternative models for the renovation of inner areas of Lübeck') *(Nachdruck)*. Projektgruppe Stadtsanierung.

Hansestadt Lübeck, der Senat der (1978), *Die Diagnose aus dem Generalverkehrsplan* ('Issues for the Traffic Master Plan').

Hansestadt Lübeck, der Senat der (1980), *Konzeption des zukunftigen Verkehrssystems* ('Models for future traffic networks').

Hansestadt Lübeck, der Senat der (1982), *Verkehrsplanung Innenstadt* ('Traffic Planning in inner Lübeck').

Hansestadt Lübeck, der Senat der (1984), *Rahmenplanung Innenstadt* ('Structure planning for inner Lübeck').

Hansestadt Lübeck, der Senat der (1984), *Rahmenplanung Innenstadt* ('Structure planning for inner Lübeck'), *Fortschreibung '84*.

Hansestadt Lübeck, der Senat der (1985), *Radwegeplanung* ('Planning for cycleways').

Hass-Klau, C. (1982), 'New transport technologies in the Federal Republic of Germany', *Built Environment*, 8, 190–7.

Hayward, J. and Watson, M. (1975), *Planning, politics and public policy, the British, French and Italian experience*, Cambridge University Press, Cambridge.

Hessischer Minister des Innern (1982), *Der Burger wirkt mit* ('Citizen participation'), Wiesbaden.

Hetzenecker, B.R., Haak, A. and Jencke, P. (1983), *Planungsstudie zum Einsatz von automatisch quergeführten Bussen in Regensburg* ('Planning study on the employment of automatic transversally operating buses in Regensburg'), Regensburger Verkehrsbetriebe GmbH.

Hibbs, J. (1985), *Regulation: an international study of bus and coach licensing*, Transport Publishing Projects, Cardiff.

Hollmann, H. (1977), 'Grenzüberschreitende Landesplanung Bremen/Niedersachsen' ('Cross-boundary State planning Bremen/Lower Saxony'), *Raumforschung Raumordnung*, 35, 218–24.

Howard, E.B. and Davies, R.L. (1986), *Contemporary change in Newcastle city centre and the impact of the Metro*, discussion paper no. 77, Centre for Urban and Regional Development Studies, University of Newcastle upon Tyne.

Husain, M.S. (1980), 'Office development in Hamburg: the City-Nord project', *Geography* 65, 131–4.

Institut nationale de la statistique et des études économiques (1981), *Annuaire statistique de la France* ('Annual statistical report on France'), Paris.

International Railway Journal (1985), 'RATP tries low capacity transit systems in Paris', November 1985, 25, 45–8.

International Road Federation (1984), *World Transport Statistics 1979–83*, Geneva and Washington.

Jobert, B. (1975), 'Urban planning and political institutions: an essay in comparison', in Hayward, J. and Watson, M. (Eds), *Planning, politics and public policy, the British, French and Italian experience*, Cambridge University Press, Cambridge.

Kimminich, O. (1981), 'Public participation in the Federal Republic of Germany', *Town Planning Review*, 52: 247–9.

Kinsey, J. (1979), 'The Algerian movement to Greater Marseille', *Geography*, 64, 338–41.

Konukiewitz, M. and Wollmann, H. (1982), 'Physical planning in a Federal System: the case of West Germany', in McKay, D.H. (Ed) *Planning and politics in Western Europe*, Macmillan, London and Basingstoke.

Kruger, T. Rathmann, P and Utech, J. (1972), 'Das Hamburger Dichtemodell' ('The Hamburg Density Model'), *Städtbauwelt*, 36.

Kunzmann, K.R. (1984), 'The Federal Republic of Germany', Chapter 2 of Williams, R.H. (Ed), *Planning in Europe, urban and regional planning in the EEC*, George Allen & Unwin, London.

le Guénédal, N. (1981), 'Besançon: aller toujours plus loin', *Transport Public*, 785, July/August.

Lichfield, N., Kettle P. and Whitbread, M. (1975), *Evaluation in the planning process*, Pergamon, Oxford.

Lille, Communauté Urbaine de (1984a), *Découvrez votre métro* ('Discover your metro').

Lille, Communauté Urbaine de (1984b), *En direct du métro* ('Straight from the metro'), 13, Lille.

Ling, A. (1967), *Runcorn New Town*, Runcorn Development Corporation.

Lissarague, P. (1984), 'La politique des déplacements à Marseille' ('Policies towards journeys in Marseille'), *Transports, Urbanisme, Planification*, 1, 5–19, Centre d'Etudes des Transports Urbains, Ministère des Transports/ Ministère de l'Urbanisme et du Logement.

London Regional Transport (1985), *Report and accounts 15 months ended 31st March 1985*.

London Regional Transport (1986), *Annual Report and Accounts 1985/86*.

London Regional Transport Act (1984), Chapter 32, HMSO, London.

London Transport (1973), *A report on the traffic implications of the Victoria Line north of Victoria*.

London Transport (1984), *London transport statistics as at 1st January 1984*.

London Transport/GLC (1982), *Travel Diary Panel*, October/November.

London Underground Limited (1986), *London's Underground – a brief history*, Press and Public Relations, London Underground Ltd.

Lyon, Agence d'Urbanisme de la Communauté Urbaine de (1983), *Atlas communale de la région lyonnaise. Recensement de la population* ('Atlas of the Lyon Region by commune. Census of population').

Mackett, R.L. (1984), *The impact of transport policy on the city*, TRRL supplementary report 821, Department of the Environment/Department of Transport, Crowthorne.

Malabry, M. (1985), 'Routes, Val et Centre Commercial' ('Routes, the Val métro and the commercial centre'), *Transports, Urbanisme, Planification*, 5, 85–104, Centre d'Etudes des Transports Urbains, Ministère des Transports/ Ministère de l'Urbanisme et du Logement.

Marseille, Ville de (1983a), *Marseille Informations. Une deuxieme ligne de métro pour 1984*, 141, 142, February/March.

Marseille, Ville de (1983b), *Trente années de dynamisme municipal* ('Thirty years of municipal achievement').

Mazzella, P. (1981), *Contribution au colloque Urbanisme et Transports*, FNAU, Vè rencontre, Lyon 14 et 15 Octobre 1981, Agence d'Urbanisme de l'Agglomération Marseillaise.

Mazella, P. (1984), 'Deplacement et vie de quartier a Marseille' ('Journeys and local community spirit in Marseille'), *Transports, Urbanisme, Planification*, 4, 5–13, Centre d'Etudes des Transports Urbains, Ministère des Transports/Ministère de l'Urbanisme et du Logement.

McKay, D.H. (Ed) (1982), *Planning and politics in Western Europe*, Macmillan, London and Basingstoke.

Mény, Y. (1982), 'Urban planning in France: dirigisme and pragmatism 1945–80', in McKay, D. (Ed), *Planning and politics in Western Europe*, Macmillan, London and Basingstoke.

Minvielle, E. (1985), 'Besoin de financement et évolution de l'usage des réseaux de transport collectif de province' ('The need for finance and the evolution of urban transport in provincial towns'), *Transports, Urbanisme, Planification*, 5, 161–70, Centre d'Etudes des Transports Urbains, Ministère des Transports/Ministère de l'Urbanisme et du Logement.

Mogridge, M.J.H. (1985), 'Transport, land use and energy interaction', *Urban Studies*, 22, 481–92.

Monheim, R. (1974), *Bericht uber das Forschungsprojekt MPPRS*, BM Bau.

Monheim, R. (1980), *Fussgängerbereiche und Fussgängerverkehr in Stadtzentren*, Bonner Geographische Abhandlungen, Bonn.

Moseley, M.J. (1980), 'Strategic planning in the Paris agglomeration in the 1960s and 1970s', *Geoforum*, 11, 362–81.

Michael, R. (1979), 'Metropolitan development concepts and planning policies in West Germany', *Town Planning Review*, 50: 287–312.

Ministère de l'Environnement et du Cadre du Vie (Service de l'Information) (1979), *Environnement, Aménagement et Urbanisme* ('Environment, Land Management and Town Planning), 2nd ed. Paris.

Ministère de l'Environnement et du Cadre du Vie/ Ministère des Transports (1979), *Etudes de suivi des ouvertures des métros de Lyon et Marseille* ('Studies following the openings of the Lyon and Marseille métros'), Paris.

Muthesius, T. (1982), 'Höhere Parkgebühren in den Städten' ('Higher urban parking charges'), *Der Stadtetag*, 9.

Nash, C.A. (1985), 'Policies towards suburban rail services in Britain and the Federal Republic of Germany: a comparison', *Transport Review*, 5, 269–82.

Niblett, G.R. (1972), 'Bus rapid transit', *Traffic Engineering and Control*, 14, 30–4.

Nivollet, A. (1982), 'La décentralisation', *Les Cahiers Français*, 204, 1–80.

Ogden, P.E. (1985), 'Counterurbanisation in France: the results of the 1982 Population Census', *Geography*, 70, 24–35.

Paris, Ville de (1974), *20 ans de transformation de Paris 1954–74*, Association Universitaire de Récherches Geographiques et Cartographiques.

Pelliard, P. (1983), 'L'organisation des déplacements urbains: les grands choix concernant la conception et l'utilisation de la voirie et les espaces publics' ('The organization of urban journeys: the main choices concerning the design and utilization of the transport network and public land) in Ministère des Transports/Ministère de l'Urbanisme et du Logement, *Transports, Urbanisme, Planification*, Paris.

Plagnol, M. (1984), 'Dernier né de la famille des métros: le métro de Lille a six mois' ('The youngest in the métro family: the Lille métro is six months old'), *Revue Générale des Chemins de Fer*, February, 57–62.

Potter, S., (1976), *Transport and new towns. The transport assumptions underlying the design of Britain's new towns 1946–76*, Open University, Milton Keynes.

Reeds, J. (1986), 'Reaching the crossroads in tramway development', *Surveyor*, 18/25th December, 12–16.

Régie Autonome des Transports Parisiens (1977), *Des couloirs reservés aux autobus* ('Bus only lanes'), Paris.

Régie Autonome des Transports Parisiens (1981a), *L'organization des transports dans la région de l'Ile* ('The organisation of transport in the region Ile de France'), Paris.

Régie Autonome des Transports Parisiens (1981b), *Le régime financier de la RATP* ('The financial governance of the RATP'), Paris.

Régie Autonome des Transports Parisiens (1983a), *La Régie Autonome des Transports Parisiens* ('The Paris Public Transport Authority'), Paris.

Régie Autonome des Transports Parisiens (1983b), *Tarification des réseaux de la RATP* ('Fare structure of the RATP network'), Direction du développement, Service du développement commercial, Paris.

Régie Autonome des Transports Parisiens (1983c), *La RATP en bref* ('The RATP in brief'), Paris.

Régie Autonome des Transports Parisiens (1983d), *RATP 1983*, Paris.

Régie Autonome des Transports Parisiens (1984a), *Rapport annuel* ('Annual report') 1983, Paris.

Régie Autonome des Transports Parisiens (1984b), *Plan d'Entreprise 1984–1988*, Paris.

Régie des Transports de Marseille (1984a), *Notice 1984*, Marseille.

Régie des Transports de Marseille (1984b), *Structures. L'organisation de la Régie des Transports de Marseille*, Marseille.

Reichert, H. and Remond, J-D (1981a), 'Lectures de la ville', *Cahiers Français*, 203, 2–5.

Reichert, H. and Remond, J-D (1981b), 'Pathologie de la grande ville', *Cahiers Français*, 203, 66–9.

Riffault, M-C. (1981), 'Une métropole d'équilibre: Bordeaux', *Cahiers Français*, 203, 38–40.

Rigby, J.P. (1980), *Public transport planning in the shire counties*, Oxford Polytechnic, Department of Town Planning, working paper number 46, Oxford.

Roberts, J. (1981), *Pedestrian precincts in Britain*, Transport and Environmental Studies, London.

Rose, R. (Ed) (1974), *The management of urban change in Britain and Germany*, Sage, London and Beverly Hills.

Rosner, R. (1975), 'Town and regional planning in Germany', *The Planner*, 61: 375–8.

Sanson, H. (1984), 'Pour l'organisation d'un pole secondaire autour du terminal métro à Caluire (secteur nord de l'agglomération Lyonnaise) (Development of a secondary centre around the métro terminal at Caluire (in the north of Greater Lyon)', *Transports, Urbanisme, Planification*, 1,

37–54, Centre d'Etudes des Transport Urbains, Ministère des Transports/ Ministère de l'Urbanisme et du Logement.

Sanson, H. (1985), 'Pratiques d'échanges voitures particulières – transports collectifs et activités commerciales (commercial activity at points of change between private and public transport)', *Transports, Urbanisme, Planification*, 5, 105–16, Centre d'Etudes des Transport Urbains, Ministère des Transports/Ministère de l'Urbanisme et du Logement.

Scarman, Lord (1986), 'Injustice in the cities', *New Society*, 75, 14th February, 286–7.

Schaechterle, K., Stengel, W. and Heck, H.M. (1984), 'Extents and causes of shifts in traffic (new traffic) when new high speed urban railway lines are opened', *Urban Transport Research*, 34, Federal Minister of Transport, West Germany.

Scott, D. (1983), 'Regional planning and the modernisation of the West German transport system', *Geography*, 68: 266–71.

Scottish Development Department (1968), *The Central Borders – a plan for expansion*, HMSO, Edinburgh.

SETRA Division urbaine (1974), *Urbanisme et transport: les critères d'accessibilité et de développement urbain* ('Town planning and transport: accessibility criteria and urban development'), Ministère de l'Equipment, Paris.

Simpson, B.J. (1983), *Site costs in housing development*, Construction Press, London and New York.

Simpson, B.J. and Purdy, M.T. (1984), *Housing on sloping sites: A design guide*, Construction Press, London and New York.

Simpson, B.J. (1985), *Quantitative methods for planning and urban studies*, Gower, Aldershot.

Simpson, B.J. (1986), 'Driving back life into the city centres', *Surveyor*, 27th November, 18–20

Simpson, B.J. (1987), *Planning and public transport in Great Britain, France and West Germany*, Longman, London and New York.

Smith, B.A. (1977), 'Central area development the German style', *Town & Country Planning*, July/August.

Société d'Economie Mixte du Métropolitain de l'Agglomération Lyonnaise (1983), *Le métro lyonnais*, Société d'Economie Mixte du Métropolitaine de l'Agglomération Lyonnaise, Lyon.

Société d'Economie Mixte du Métropolitain de l'Agglomération Lyonnaise (1984), *Rapport et perspectives d'activites* ('Report on activities'), Lyon.

Société du Métro de Marseille (1983), *Le métro de Marseille*, Marseille.

Société du Métro de Marseille (1985), *Le métro de Marseille. La 2è ligne*, Marseille.

Société Lyonnaise des Transports en Commun (1983), *Le métro de Lyon*, Lyon.

Société Lyonnaise des Transports en Commun (1984), *Le réseau bus:métro de l'agglomération Lyonnaise*, Lyon.

Stadt Frankfurt am Main (1976), *Model für ein kreuzungsarm vernetztes Ring – bzw. Mehrkreissystem zur Fuhrung des Kfz-Hauptverkehrs in der Frankfurter City ab 1985* ('Model for a multiple interlinking radial system to guide the main road network in the City of Frankfurt up to 1985').

Stadt Frankfurt am Main (1980a), *Veränderung von Wohnflächenstandards und ihre Auswirkungen auf die raumliche Entwicklungsplanung in Frankfurt* ('Change in house building density standards and their effects on spatial development in Frankfurt'), Amt für Kommunale Gesamtentwicklung und Stadtplanung, Frankfurt.

Stadt Frankfurt am Main (1980b), *Flächennutzungsplan '80* ('Land use proposals plan '80'), Amt für kommunale Gesamtentwicklung und Stadtplanung.

Stadt Frankfurt am Main (1984), *Freiflächenentwicklungsplan* ('Open space development plan') *Landschaftsplan* ('Landscape plan'), Amt für kommunale Gesamtentwicklung und Stadtplanung.

Statistisches Bundesamt (1987), *Statistisches Jahrbuch für die Bundesrepublik Deutschland 1987*, Wiesbaden.

Stewart, J.R. (1976), *User response to pedestrianised shopping streets*, Research Memorandum 73, Centre for Urban and Regional Studies, Birmingham University.

Syndicat des Transports en Commun de la Région Lyonnaise (1979), *Document d'étude du Syndicat des Transports en Commun de la Région Lyonnaise sur les extensions du réseau de transport en site propre Lyonnais pour les dix prochaines années 1980–1990* ('Lyon Regional Transport Authority study on the Lyon fixed track transport network for the next ten years 1980–1990'), Lyon.

Syndicat des Transports en Commun de la Région Lyonnaise (1984a), *Le réseau bus:métro de l'agglomération Lyonnaise* ('The bus:metro network of the Lyon conurbation'), Lyon.

Syndicat des Transports en Commun de la Région Lyonnaise (1984b), *Rapport d'activité 1983* ('Annual report 1983'), Lyon.

Syndicat des Transport en Commun de la Région Lyonnaise (1984c), *L'agglomération Lyonnais et les transports de personnes. Memento statistique* ('The Lyon conurbation and passenger transport. Statistical memorandum'), Lyon.

Syndicat des Transports en Commun de la Région Lyonnaise/ Agence d'Urbanisme de la Communauté Urbaine de Lyon (1983), *Les transports en commun et l'urbanisation périphérique de l'agglomération Lyonnaise. Le diagnostic. Une coordination difficile* ('Public transport and peripheral urbanisation of the Lyon conurbation. The diagnosis. A difficulty of coordination'), *1984 Propositions, actualisation du diagnostic, propositions et recommandations* ('Propositions, realisation of the diagnosis, proposals and recommendations'), Lyon.

Szumeluk, K. (1968), *Central Place Theory: a review*, Working paper number 2, Centre for Environmental Studies, London.

Topp, H.H. (1985), 'Developments in public transport in the Federal Republic of Germany', *Transportation Quarterly*, 39: 407–27.

Tournier, G., Laferrère, M. and Thomas, F. (1973), *Lyonnais, Beaujolais, Forez, Vivarais*, Larousse, Paris

Tuppen, J. (1980), 'Public transport in France: the development and extension of the metro', *Geography*, 65, 127–30.

Tuppen, J. (1986), 'Core-periphery in metropolitan development and planning: socio-economic change in Lyon since 1960', *Geoforum*, 17, 1–37.

Tym, R. and Partners (1983), *West Midlands Rapid Transit Study: planning and development*, West Midlands County Council, Birmingham.

Tyne and Wear County Council (1985), *Structure plan annual report 1984*, Newcastle upon Tyne.

Utech, J. (1982), 'Das Hamburger Dichtemodell 1980 und seine Wirkungsmöglichkeiten auf das Schnellbahn Fahrgastaufkommen' ('The Hamburg density model 1980 and its possible effects on the use of the railways by passengers'), *Verkehr und Technik*, 9.

Vincent, R.A., Layfield, R.E. and Bardsley, M.D. (1976), *Runcorn busway study*, TRRL Report 697, Department of the Environment/Department of Transport, Crowthorne.

von Klitzing (1974), 'The description of urban development in terms of land use in Germany', in Rose, R. (ed), *The management of urban change in Britain and Germany*, Sage, London and Beverly Hills.

Waldmann, R. and Ferrand, J. (1977), 'Le métro: expression et moyen pour l'urbanisme de l'agglomération Lyonnaise' ('The métro, symbol and means for planning the Lyon conurbation'), in *Le métro Lyonnais et la qualité de la vie* ('The Lyon métro and the environment'), Technica, 394, Lyon.

Wardrop, J.G. (1975), *Review of traffic capacity with special reference to central urban areas*, TRRL SR 231, Department of the Environment/ Department of Transport, Crowthorne.

Watel, M. (1984), *Déplacements, transports et commerce. La presqu'île 1975–80* ('Journeys, transport and commerce. The peninsula 1975–80'), Journée Nationale d'étude 6/12/84, CETUR, CETE, CECOD, Lyon.

Watel, M. (1985), 'Activites commerciales de la presqu'île de Lyon' ('Commercial activities in the centre of Lyon'), *Transports, Urbanisme, Planification*, 5, 53–71, Centre d'Etudes des Transports Urbains, Ministère des Transports,/Ministère de l'Urbanisme et du Logement.

Webster, F.V., Bly, P.H., Johnson, R.H., Paulley, N. and Dasgupta, M. (1985), 'Evolution des structures de déplacements urbains' ('Evolution of the composition of urban journeys'), *Transports, Urbanisme, Planification*, 4, 135–48, Centre d'Etudes des Transports Urbains, Ministère des Transport/ Ministère de l'Urbanisme et du Logement.

West Midlands County Council (1980), *County Structure Plan written statement*, Birmingham.

West Midlands County Council (1984), *Transport policies and programme 1984 submission*, Birmingham.

West Midlands County Council/West Midlands Passenger Transport Executive (1984), *Rapid Transit for the West Midlands, final report*. Birmingham.

West Midlands Passenger Transport Executive (1984a), *Annual statistical report 1983–84*, Market Research and Management Services Unit, Birmingham.

West Midlands Passenger Transport Executive (1984b), *Three year transport plan*, Birmingham.

Wild, T. (ed) (1983), *Urban and rural change in West Germany*, Croom Helm.

Williams, R. (1978), 'Urban Planning in Federal Germany', *The Planner*, 64, 46–7.

Williams, R.H. (ed) (1984), *Planning in Europe. Urban and Regional Planning in the EEC*, George Allen & Unwin, London.

Wilson, H. and Womersley, L. (1966), *Redditch new town: report on planning proposals*, Redditch Development Corporation.

Wilson, I.B. (1983), 'Preparation of local plans in France', *Town Planning Review*, 54: 155–73.

Wilson, T.K. and Bell, M.C. (1985), 'Transport coordination and integration in West Germany', *Highways and Transportation*, 32: 5–12.

Wistrich, E. (1983), *The politics of transport*, Longman, London and New York.

Index

197

For Product Safety Concerns and Information please contact our EU
representative GPSR@taylorandfrancis.com
Taylor & Francis Verlag GmbH, Kaufingerstraße 24, 80331 München, Germany